Microsoft Power Pages in Action

Accelerate your low-code journey with functional-rich web apps using Power Pages

Faisal Hussona

Microsoft Power Pages in Action

Group Product Manager: Aaron Tanna

Publishing Product Manager: Uzma Sheerin

Senior Content Development Editor: Rosal Colaco

Book Project Manager: Deeksha Thakkar

Technical Editor: Rajdeep Chakraborty

Copy Editor: Safis Editing

Indexer: Rekha Nair

Production Designer: Prafulla Nikalje

DevRel Marketing Coordinator: Deepak Kumar and Mayank Kumar

First published: June 2024

Production reference: 1140624

Published by Packt Publishing Ltd.

Grosvenor House

11 St Paul's Square

Birmingham

B3 1RB, UK

ISBN 978-1-83763-045-5

www.packtpub.com

Contributors

About the author

Faisal Hussona, a seasoned IT professional with over two decades of experience, specializes in developing and implementing technology solutions across diverse industries, with a particular focus on Power Platform and Dynamics 365. His successful track record includes delivering projects that emphasize promoting citizen developers and facilitating technology transfer. Throughout his 25-year career, Faisal has mentored and trained individuals, transferring knowledge and development skills to various organizations. His expertise in Power Platform and Dynamics 365 has led to successful projects, particularly in improving business processes and enhancing customer experience.

About the reviewer

Danilo Capuano is a Low Code Practice Lead and Office Manager at Agic Technology. He is 6x Microsoft Certified Trainer and 1x Microsoft MVP.

Danilo helps companies scale innovation with Microsoft Power Platform and Dynamics 365 CRM. He is active in the Microsoft community in Italy, being a group leader for Power Apps User Group Italia, Power Pages User Group Italia, and Copilot User Group Italia. He has previously worked as a technical reviewer on several other titles, including Microsoft Power Platform Enterprise Architecture, Fundamentals of CRM with Dynamics 365 and Power Platform, Learn Microsoft Power Apps, Mastering Microsoft Dynamics NAV 2016, Microsoft Dynamics NAV 7 Programming Cookbook, Microsoft Dynamics NAV 2013 Application Design and Programming Microsoft Dynamics NAV 2015.

Table of Contents

6

Basic Forms, Lists, and Web Pages   103

7

JavaScript and jQuery   141

8

Web Templates and Liquid   157

9

Workflow Automation   179

10

Power Pages and Cloud Flows   195

11

Charts, Dashboards, and Power BI   211

12

REST Integration   231

13

Creating a PDF File from Dataverse   255

14

15

Preface

Welcome to *"Microsoft Power Pages in Action."* This book is more than just a beginner's guide—it's your comprehensive roadmap to building and customizing functional and rich web apps effortlessly. Whether you're a citizen developer just starting out or an experienced professional looking to enhance your skills, this book is designed to equip you with a practical understanding of the Power Pages environment through real-world examples and step-by-step tutorials.

Our journey begins with the fundamentals of creating data tables and forms. From there, you'll learn how to add and customize web pages, integrating advanced techniques to seamlessly enhance your applications. By the end of this book, you'll have mastered the art of creating responsive, feature-rich web pages that not only meet but exceed your needs.

Throughout the chapters, you'll explore how to automate your web apps, enhance virtual agent bots, and implement a web app for incident management systems. These skills will enable you to improve user experience and data quality during the insertion of web app data. You'll also discover how to craft portals and websites using various functionalities and layouts, harnessing the capabilities of Microsoft technologies like Dataverse, Power Automate, Power Pages Studio, and Visual Studio Code.

In "Microsoft Power Pages in Action," you will learn how to fully leverage Power Pages to create interactive and dynamic web pages and portals. You will be guided on establishing secure portals with robust configurations, ensuring data integrity and user protection throughout your applications. You will delve into using ChatGPT to enhance your designs and website code development. Additionally, you will discover how to implement responsive designs, making your apps adaptable across various devices. The integration of Power Pages with other Microsoft technologies will be covered extensively, enabling you to create cohesive and powerful digital solutions. You will also gain proficiency in developing automation flows to integrate external services and enhanced web page functionality.

Packed with essential features, this book will help you build custom web apps effortlessly using Power Pages' low-code platform. You will learn to create responsive, intuitive interfaces with practical design tips, enhancing your web apps with advanced features and automation.

As you embark on this journey, embrace the collaborative spirit of the Power Platform community. Our goal is to inspire you to explore new possibilities and drive meaningful change within your organization. By mastering Microsoft Power Pages, you'll become a catalyst for innovation, helping your business navigate the digital era with agility and confidence.

Join us on this exciting journey of discovery and transformation. Welcome to *"Microsoft Power Pages in Action."*

Who this book is for

This book is the ultimate guide for citizen developers aiming to build functionally rich and responsive web apps with Power Pages. With clear and concise guidance, it caters to both beginners and experienced developers, offering practical insights into every aspect of web app development. Whether you're new to the field or seeking advanced techniques, this book equips you with the skills you need to create powerful and user-friendly web apps tailored to your unique needs.

What this book covers

Chapter 1, Modernizing Rob the Builder's Business with Power Pages, will cover empowering Rob the Builder's transformation, Power Pages – a paradigm shift, Setting the stage – tenancy and Power Pages, and navigating the developer's landscape. We introduce Rob, Brenda, and Sarah as the main characters in this book's story.

Chapter 2, Power Pages Design Studio, will explore the exciting journey of Sarah and Brenda as they dive into the world of the Power Pages design studio to transform the digital presence of Rob the Builder.

Chapter 3, Power Pages Studio – Styling and Themes, aims to guide Sarah through the intricate process of aligning the website with the new branding guidelines, employing advanced styling techniques, and ensuring that the final product resonates with Rob the Builder's newly established brand identity.

Chapter 4, Dataverse Tables and Forms, will help you learn how Sarah utilizes Power Pages Design Studio and Dataverse solutions to implement tables and forms. You will also understand the step-by-step process of translating a conceptual design into a functioning system.

Chapter 5, Table Permissions and Security, will cover exploring how table permissions enable access to Dataverse records in Power Pages, with role-based security access, configuring table permissions, child access permissions, and introduction to the architecture of table permissions.

Chapter 6, Basic Forms, Lists, and Web Pages, in this chapter, Sarah's focus will be on creating three essential forms: insert, edit, and read-only forms. Additionally, Sarah will design three corresponding web pages for each form, as well as a list page that provides convenient access to these forms as a hub or landing page.

Chapter 7, JavaScript and jQuery, helps Sarah explore the advice she received and the JavaScript and jQuery code she utilized to fulfill Brenda's requests. Sarah will delve into the world of JavaScript and jQuery to implement dynamic form interactions.

Chapter 8, Web Templates and Liquid, helps Sarah uncover the dynamic duo of web templates and Liquid. This chapter helps Sarah understand and leverage the combination of web templates and Liquid to create more engaging and interactive Power Pages experiences.

Chapter 9, Workflow Automation, helps Sarah explore the realm of processes and automation within the Power Pages framework. By the end of this chapter, Sarah will be able to understand Dataverse Workflow in Power Pages.

Chapter 10, Power Pages and Cloud Flows, covers the key features of Power Automate in Power Pages, integration with Power Automate, use case – automating email notifications for timesheet approvals, cloud flow triggered by Dataverse, and cloud flow integrated with Power Pages.

Chapter 11, Charts, Dashboards, and Power BI, explores the world of data-driven visualizations in Power Pages. Sarah will delve into the concepts and functionalities of charts, dashboards, and Power BI integration, providing a comprehensive understanding of their role in presenting data-driven insights.

Chapter 12, REST Integration, helps Sarah to design a solution to integrate the website with the Xero accounting system, design web pages and a basic form to support the integration, design a page to handle a two-step login process for REST authentication, design and code a web template to handle the integration and JavaScript, implement a Power Automate cloud flow to make HTTP requests to post invoices to the accounting system, and implement JavaScript to call the Power Automate flow.

Chapter 13, Creating a PDF File from Dataverse, addresses client requirements for generating PDF invoices for customer accessibility and email attachments. We will also discuss design considerations to ensure seamless PDF generation. Additionally, we will cover the utilization of Dataverse file fields for storing PDFs, addressing challenges such as user access and file replacement. The chapter will also delve into using Office Word templates to streamline PDF generation and leveraging Power Automate cloud flows to automate PDF generation. Finally, we will explore OneDrive integration for converting Word templates to PDFs, taking into consideration cost-effectiveness and accessibility.

Chapter 14, Modal Windows, helps Sarah explore the concept of modal windows within Power Pages. Sarah is going to further explore how she can work with modal windows to deliver improved UX.

Chapter 15, Enhancing Development with ChatGPT, helps Sarah explore how AI-powered development tools, specifically ChatGPT and Copilot, have revolutionized her development process. These tools have become invaluable allies, providing guidance, generating code snippets, and helping her debug issues more efficiently than ever before.

To get the most out of this book

Software/hardware covered in the book	Operating system requirements
Power Pages studio	Windows, macOS, or Linux
Power Pages Management	Windows, macOS, or Linux
Microsoft Visual Studio Code	Windows, macOS, or Linux
Power Apps CLI	Windows, macOS

If you are using the digital version of this book, we advise you to type the code yourself or access the code from the book's GitHub repository (a link is available in the next section). Doing so will help you avoid any potential errors related to the copying and pasting of code.

Download the example code files

You can download the example code files for this book from GitHub at `https://github.com/PacktPublishing/Microsoft-Power-Pages-In-Action`. If there's an update to the code, it will be updated in the GitHub repository.

We also have other code bundles from our rich catalog of books and videos available at `https://github.com/PacktPublishing/`. Check them out!

Conventions used

There are a number of text conventions used throughout this book.

`Code in text`: Indicates code words in text, database table names, folder names, filenames, file extensions, pathnames, dummy URLs, user input, and Twitter handles. Here is an example: "In a Microsoft Power Pages web template using Liquid and jQuery, I have encountered an error in the JavaScript function `GetExternalData()`."

A block of code is set as follows:

```
function fetchData() {
    fetch('https://api.example.com/data')
        .then(response => response.json())
        .then(data => {
            // Process data
            console.log(data);
        })
        .catch(error => console.error('Error:', error));
}
```

Any command-line input or output is written as follows:

```
async function fetchData() {
    try {
        const response = await fetch('https://api.example.com/data');
        const data = await response.json();
        console.log(data);
    } catch (error) {
        console.error('Error:', error);
    }
}
```

Bold: Indicates a new term, an important word, or words that you see onscreen. For instance, words in menus or dialog boxes appear in **bold**. Here is an example: "Sarah selected **Sections** from Copilot and tried to create an HTML table within a section."

> Tips or important notes
> Appear like this.

Get in touch

Feedback from our readers is always welcome.

General feedback: If you have questions about any aspect of this book, email us at customercare@packtpub.com and mention the book title in the subject of your message.

Errata: Although we have taken every care to ensure the accuracy of our content, mistakes do happen. If you have found a mistake in this book, we would be grateful if you would report this to us. Please visit www.packtpub.com/support/errata and fill in the form.

Piracy: If you come across any illegal copies of our works in any form on the internet, we would be grateful if you would provide us with the location address or website name. Please contact us at copyright@packt.com with a link to the material.

If you are interested in becoming an author: If there is a topic that you have expertise in and you are interested in either writing or contributing to a book, please visit authors.packtpub.com.

Share Your Thoughts

Once you've read *Microsoft Power Pages in Action*, we'd love to hear your thoughts! Scan the QR code below to go straight to the Amazon review page for this book and share your feedback.

https://packt.link/r/1837630453

Your review is important to us and the tech community and will help us make sure we're delivering excellent quality content.

Download a free PDF copy of this book

Thanks for purchasing this book!

Do you like to read on the go but are unable to carry your print books everywhere?

Is your eBook purchase not compatible with the device of your choice?

Don't worry, now with every Packt book you get a DRM-free PDF version of that book at no cost.

Read anywhere, any place, on any device. Search, copy, and paste code from your favorite technical books directly into your application.

The perks don't stop there, you can get exclusive access to discounts, newsletters, and great free content in your inbox daily

Follow these simple steps to get the benefits:

1. Scan the QR code or visit the link below

https://packt.link/free-ebook/9781837630455

2. Submit your proof of purchase
3. That's it! We'll send your free PDF and other benefits to your email directly

Modernizing Rob the Builder's Business with Power Pages

In this book, we will follow the journey of transforming Rob the Builder's business with Microsoft Power Pages. Through this, you will become acquainted with the key players who shape this narrative:

- **Rob**, the determined owner of Rob the Builder, an experienced builder and engineer, overseeing a thriving construction company

- **Brenda**, Rob's business partner and the company's backbone

- **Sarah**, Brenda's tech-savvy niece, a recent graduate with a budding passion for technology, especially the Power Platform

Together, they seek to modernize the construction business and bridge technology gaps. This chapter will cover the paradigm shift brought by Power Pages to setting up the tenancy and guiding Sarah through the installation process.

We'll cover the following topics in this chapter:

- Empowering Rob the Builder's transformation

- Power Pages – a paradigm shift

- Setting the stage – tenancy and Power Pages

- Navigating the developer's landscape

Empowering Rob the Builder's transformation

In a quiet corner of the bustling city, Rob was at the helm of Rob the Builder, a modest yet thriving building construction company. With a dedicated team that included his wife Brenda and several family members, they had successfully completed numerous projects. However, behind their success stories lurked an undercurrent of inefficiency and missed opportunities. The tale of Rob the Builder is not just about bricks and mortar; it's a narrative of evolution, innovation, and the power of technology to reshape a business's trajectory.

Brenda's revelation

Behind the scenes, Brenda tirelessly managed the company's back-office operations using a patchwork of tools – Excel, Word, and dated accounting software. Invoices, remittance advice, and other documents were created using Word templates that had served them for years. While the business thrived, Brenda knew they needed to shed their outdated processes to stay relevant in a rapidly changing world.

Lost opportunities and health and safety compliance

The turning point came when Rob realized they had missed out on lucrative project opportunities due to their outdated systems. Large clients demanded stringent health and safety reporting, an area where Rob the Builder struggled. As a result, they lost credibility and potential contracts. Something had to change, and Brenda's recent discovery was their beacon of hope.

Brenda's niece, Sarah, enters the scene

Enter Sarah, Brenda's niece and a recent graduate. Armed with a smattering of HTML and JavaScript knowledge from her college web development course, Sarah had caught wind of the Power Platform during Microsoft exhibitions. Though she lacked extensive software development skills, Sarah was convinced that the Power Platform's no-code/low-code approach could be the catalyst for Rob the Builder's transformation.

Power Pages – a vision unveiled

After heartfelt discussions between Rob, Brenda, and Sarah, the decision was made to bring Sarah aboard. Brenda, motivated by a desire to modernize the business, would work side by side with Sarah to build a business system centered on the Power Pages site. Their first target is an **incident management** (**IM**) system. However, the scope extended beyond that – encompassing **customer relationship management** (**CRM**), invoicing, estimating, and even a virtual assistant to revolutionize customer interactions.

Power Pages – a paradigm shift

In the ever-evolving landscape of business and technology, adaptability is not just an advantage; it's a necessity. For Rob the Builder, a construction company grappling with outdated processes and missed opportunities, this realization opened the door to a transformative journey, empowering the business to bridge the gap between technology and its unique operational needs.

Use case – establishing and utilizing a comprehensive information system with Power Pages

Sarah starts with a use case to formalize the plan:

- **Scope**: Rob the Builder – construction company digital transformation

- **Primary actor**: Sarah

- **Stakeholders and interests**:

    - **Rob**: Aims to modernize the company's technology infrastructure to enhance operational efficiency and customer service (and consequently, profitability)

    - **Brenda**: Looks to simplify back-office operations, improving data management and customer communications

    - **Clients**: Expect a reliable and transparent digital interface for interacting with the company and monitoring project progress

- **Preconditions**:

    - The company does not currently have an established Microsoft Power Platform environment

    - Sarah has basic knowledge of HTML and JavaScript and is familiar with the Power Platform's capabilities

- Main success scenario:

    I. **Setting up the tenancy**:

        - Sarah registers for a Microsoft 365 subscription to provide the underlying infrastructure needed for the Power Platform

        - She sets up a new tenant in Microsoft 365, which includes configuring basic settings and integrating domain information

        - Sarah ensures that the appropriate Power Platform capabilities are enabled and that she and other key personnel have the necessary licenses

    II. **Creating a Power Platform environment**:

        - In the Power Platform admin center, Sarah creates a new environment specifically for Rob the Builder's operations, choosing the correct region and ensuring it includes a database for Dataverse

III.   **Developing a comprehensive information system**:

- Sarah uses Power Pages to design and develop a multi-functional website that includes the following:

- **IM system**: To log and monitor workplace incidents

- **CRM**: For managing all customer interactions and feedback

- **Project tracking module**: To provide real-time updates on construction projects to both staff and clients

- The website integrates seamlessly with Dataverse for robust data management and security

IV.   **Testing and implementing the system**:

- After developing the system, Sarah sets up a pilot phase involving selected stakeholders to test the functionalities and gather feedback

- Necessary adjustments are made based on the feedback to optimize usability and functionality

V.   **Going live and monitoring**:

- The comprehensive information system is officially launched

- Sarah monitors the system's performance and usage, making adjustments as needed to improve operations and customer satisfaction

- **Use-case extensions**:

  - If Sarah encounters issues with licensing or setup, she consults Microsoft support for guidance

  - If additional modules are required, such as supply chain tracking, these are developed and integrated during the pilot phase or subsequent updates

- **Frequency of use**:

  - Daily by staff for managing operations and customer interactions

  - Continuously by clients for accessing project information and communication tools

- **Special requirements**:

  - The system must be intuitive and accessible on various devices, including mobiles, to support on-site and remote use

  - Compliance with the **General Data Protection Regulation** (**GDPR**) and other relevant data protection regulations is essential

- **Technology and data variations list**:

  - The system is developed and managed using Power Pages and Dataverse within the established Power Platform environment

  - Potential integration of Power Automate to automate workflows and data synchronization between modules

This refined use case outlines a detailed pathway from initial setup to the deployment and operation of a comprehensive information system using Microsoft Power Pages, making it a practical blueprint for implementation.

Understanding Power Pages' role

At its core, Power Pages is a game-changer, offering a gateway to enhanced customer engagement and streamlined processes. Imagine a digital interface that seamlessly integrates with your existing systems, creating an environment where interactions are not just transactions but meaningful engagements. Power Pages is where the organization's digital presence takes center stage – from customer-facing applications to internal workflows.

Power Platform's no-code/low-code magic

Traditional software development often feels like an arcane art accessible only to a select few. Enter the Power Platform, a revolutionary approach that empowers individuals such as Sarah – with some HTML and JavaScript know-how – to create sophisticated applications without delving deep into code. Power Pages, a component of the Power Platform, embraces the concept of no-code/low-code development. This means Sarah can build functional and powerful applications with minimal coding, relying on intuitive drag-and-drop interfaces, pre-built templates, and configurable elements.

Bridging technology gaps with Power Pages

In the context of Rob the Builder, Power Pages becomes the bridge between a world of manual processes and the digital future. As Sarah embarks on developing solutions for IM, CRM, and more, Power Pages offers a conduit to fuse technology with the business's unique requirements. With Power Pages, the gap between aspirations and reality narrows significantly.

As Sarah navigates the landscape of tenancy setup, developer installations, and development best practices, Power Pages will emerge as her ally in modernizing every aspect of the business. In the next section, we will learn how to set up her tenancy.

Setting the stage – tenancy and Power Pages

In the journey to transform Rob the Builder from a realm of manual processes to a digitized operation, one of the crucial initial steps is laying the foundation through the concept of tenancy. This section will demystify the notion of tenancy and guide Sarah through the process of establishing a tenancy for Rob the Builder within the Power Pages environment.

Laying the foundation – what is a tenancy?

Before delving into the technical aspects, it's essential to grasp the concept of tenancy within the context of Power Pages. A tenancy serves as a dedicated workspace or environment within the Power Pages platform. It's here that you can craft, customize, and deploy applications tailored to specific needs. Each tenancy operates in isolation, ensuring data separation, security, and individualized experiences for users.

> **Tip**
>
> For further reading, follow the link to *Microsoft Learn*: `https://learn.microsoft.com/en-us/training/modules/administrating-power-platform-subscriptions/`

Establishing a tenancy for Rob the Builder

Sarah's first task is to create a tenancy for Rob the Builder, effectively carving out a unique realm where the company's digital transformation will unfold. The process is straightforward. Creating a new tenant for the Microsoft Power Platform, specifically for using Dataverse and Power Pages, involves a few key steps. Let's look at a detailed guide Sarah followed.

Step 1 – Registering for Microsoft 365 and setting up a new tenant

When purchasing Microsoft 365, a tenant for the organization is automatically created. A tenant represents an instance of Azure AD services, and it's where the organization's data is stored. Firstly, Sarah needs to have a Microsoft 365 subscription as it provides the underlying infrastructure for the Power Platform. Here are the steps she follows:

1. Sarah goes to the Microsoft 365 admin center (`admin.microsoft.com`).
2. Sarah chooses a plan that suits her development needs and follows the instructions to purchase and set up the subscription.
3. She logs in to the Microsoft 365 admin center.

4. Then, she completes any additional setup, such as adding a domain and configuring basic settings, as shown in *Figure 1.1*.

5. She enables **Users** and **Assign Licenses**.

6. To use Power Platform services, Sarah will set herself the appropriate licenses.

7. Then, she navigates to **Users | Active users** in the Microsoft 365 admin center.

8. Lastly, she adds users and assigns licenses that include Power Platform capabilities.

After following these steps, Sarah now has access to the Power Platform admin center, a hub that enables administrators to manage environments, including tenancies (`https://admin.powerplatform.microsoft.com/`):

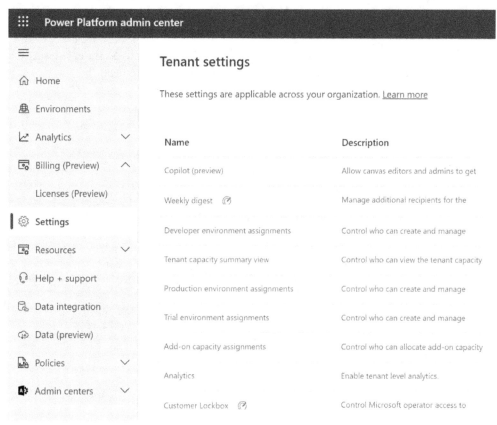

Figure 1.1 – Accessing the Power Platform admin center to begin the process of establishing a tenancy

Step 2 – Creating new environments

Power Platform environments are containers that house, manage, and share your business data, apps, and flows.

Firstly, to create a new environment within the Power Platform admin center, Sarah clicks on **Environments** and then **+ New**. She chooses a region closest to her users to ensure data can be transferred quickly.

Sarah will create two environments:

- A sandbox environment, which will later be converted to a production environment, acting as the live environment for customers.

- A development environment, allowing for tests to be made without impacting the end client. These environments are dictated by the **Type** setting (as seen in *Figure 1.2*).

Further, as she would like a database assigned to her Power Pages website, Sarah enables **Add a Dataverse data store?**. This will come in handy for future customer management:

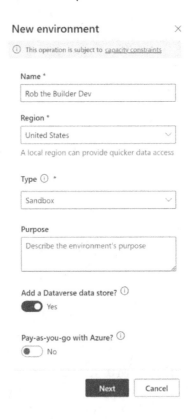

Figure 1.2 – Creating a dedicated environment with the Power Platform admin center

Configuring environment settings

Sarah can configure various settings for the environment, including data retention policies, security measures, and user access controls. Inside the newly created environment, she explores settings to configure security roles, data policies, and any integrations.

Step 3 – Power Pages setup

With the tenancy in place, it's time to venture into the Power Pages setup, where the magic truly happens. To create her first site in the new environment, Sarah will open Power Pages administration, located at `https://make.powerpages.microsoft.com/`.

Note that this is a different URL from the Power Platform admin center. Hereby, Sarah can create and manage websites. With templates to help her start websites with themes and sample pages, Sarah will browse through the templates and select one of them as her starter site. Microsoft provides a 30-day free trial for sites created, which means Sarah can try out different templates. Sarah chooses a starter layout to create an initial website to learn and start her development. Sarah enters the site name and address and creates the new site:

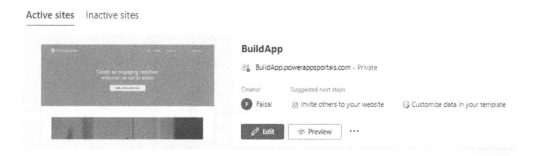

Figure 1.3 – Power Pages administration edit to open the studio

Step 4 – Power Pages studio

Once the new website is created, there is an **Edit** website link (see *Figure 1.3*) that opens the Power Pages studio. In later chapters, you will see how Sarah uses this to manage and edit the website.

Navigating the developer's landscape

In this section, Sarah will explore the Power Platform. She'll also look at the tools for Power Pages development, and configure the environment for seamless collaboration.

Exploring the Power Platform

Sarah should navigate the Power Platform, where she'll find Power Pages, Power Automate, and Power BI, forming a trifecta of no-code/low-code magic.

Browse to `https://make.powerapps.com/`, and you'll see the following menu on the left-hand side of your welcome screen:

Figure 1.4 – Sarah's journey as a developer involves exploring the capabilities of the Power Platform

Clicking on **Apps** will show all the apps installed, as shown in *Figure 1.5*, with **Power Pages Management** being the critical tool that allows her to manage her forms and their behavior, with a rich set of codeless functionality available through the management tool:

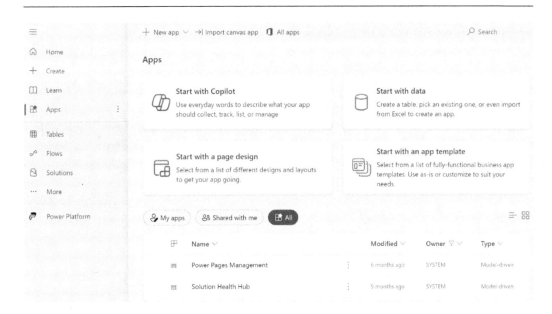

Figure 1.5 – Apps view in the Power Platform

By clicking on the **Tables** tab shown in *Figure 1.5*, Sarah can access Dataverse to allow quick access to table data, which will help her in testing her pages during development, and also to create ad hoc views while working with and viewing reference data such as often-used record types on every web page that she will work on in her projects. With the **Solutions** tab shown in *Figure 1.5*, Sarah will learn that though **solutions** are also conveniently placed in her primary tool, the Power Pages studio, there are certain features of working with solutions that are only found here.

Before Sarah dives into the creative process, she needs to get familiar with the right tools at her disposal.

Tools for Power Pages development

We will list the tools that Sarah will utilize for her development, and in the next chapter, we will learn how to use these tools. Accessing these tools sets the stage for her digital craftsmanship.

Power Pages studio

The Power Pages studio is the cornerstone tool for Sarah's web development endeavors. To access the Power Pages studio and its suite of features, Sarah navigates to `https://make.powerpages.microsoft.com/`. This platform serves as her central hub for managing websites and initiating the creation of new ones, as demonstrated earlier in this chapter. Beyond website management, the Power Pages studio offers access to essential tools such as Microsoft Visual Studio Code, Power Automate for cloud flow automation, and data for Dataverse data. Within this unified environment, Sarah can seamlessly interact with Dataverse components such as tables, views, and Dataverse forms, streamlining her development workflow and enhancing productivity.

Website deployment with solutions

Sarah will learn to use solutions to deploy her websites and their components between environments from her development environment to her production environment. Solutions are accessible via the **Solutions** tab in Power Pages home, as shown in *Figure 1.6*:

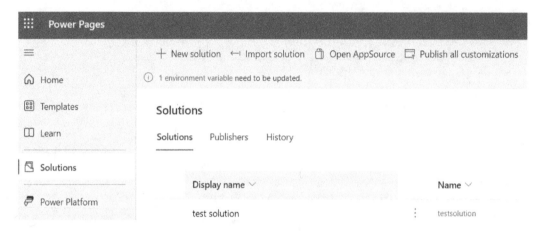

Figure 1.6 – Solutions on the Power Pages studio

Power Pages Management

Power Pages Management serves as the central hub for Sarah to oversee, organize, and maintain her Power Pages projects efficiently. It provides a comprehensive suite of tools and features tailored to streamline the management of websites created using Power Pages. **Power Pages Management** is accessed from the Power Pages studio, as shown in *Figure 1.7*:

Figure 1.7 – Accessing Power Pages Management

Here are those tools and features:

- **Project organization**: **Power Pages Management** offers robust capabilities for organizing and structuring Sarah's Power Pages projects. She can create, view, and manage multiple projects, making it easy to stay organized and keep track of her development efforts.

- **Site configuration**: Within **Power Pages Management**, Sarah can configure various aspects of her websites, including site settings, navigation menus, and user permissions.

- **Content management**: Managing content is made simple with **Power Pages Management**. Sarah can create and edit web pages, templates, and content snippets directly within the platform's intuitive interface. This is where Power Pages forms, lists pages, and buttons are configured. Sarah can open **Power Pages Management** from the Power Pages studio. Once it is open, she can browse to the **Content Snippets** tab under the **Content** section, as shown in *Figure 1.8*:

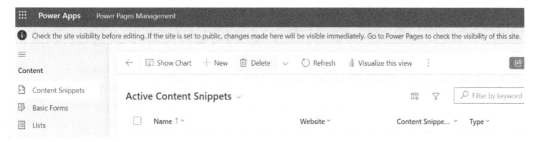

Figure 1.8 – Content Snippets on Power Pages Management

The Power Pages studio and Power Pages Management form the two most important tools that Sarah will use to do most of her work. Sarah will learn to efficiently work to keep both these open.

Microsoft Visual Studio Code

Microsoft Visual Studio Code serves as Sarah's primary workspace for various coding, customization, and integration tasks related to web development. Within this IDE, Sarah gains access to all the essential components necessary for building and managing her web pages effectively. Here's how Visual Studio Code enhances Sarah's workflow:

- **Coding environment**: Visual Studio Code provides a feature-rich coding environment where Sarah can write, edit, and debug code seamlessly. Whether it's HTML, Liquid, JavaScript, or CSS, Sarah can access and manipulate her code base with ease.

- **Customization capabilities**: With Visual Studio Code, Sarah has the flexibility to customize her coding environment according to her preferences and requirements. She can install extensions, themes, and plugins to tailor the editor to her specific needs, enhancing productivity and comfort during long coding sessions.

- **Integration with other tools**: Visual Studio Code seamlessly integrates with a wide range of tools and services, allowing Sarah to streamline her development workflow. Whether it's **version control systems** (**VCSs**) such as Git, task runners, or build automation tools, Visual Studio Code provides robust integration options, enabling Sarah to work efficiently within her preferred ecosystem.

- **AI-powered assistance with Copilot**: Visual Studio Code's integration with Copilot, an AI-powered coding assistant, further enhances Sarah's capabilities. Copilot leverages **machine learning** (**ML**) algorithms to provide code suggestions, autocomplete, and even generate code snippets based on context, making it particularly valuable for novice developers or when Sarah encounters complex coding scenarios.

- **Access code from the Power Pages studio**: The code for the website is accessed via the Power Pages studio. You can click on the **Edit code** button as shown in *Figure 1.9*:

Figure 1.9 – Accessing website code

Power Apps CLI

The Power Apps **Command-Line Interface** (**CLI**) provides a CLI for managing various components, solutions, and resources related to Power Pages development. Here's how Sarah utilizes the Power Apps CLI, especially in conjunction with Visual Studio Code, within her Power Pages projects:

- **Installation as an extension to Visual Studio Code**: Sarah seamlessly integrates the Power Apps CLI into her development environment by installing it as an extension to Visual Studio Code. This integration enhances her workflow by allowing her to access Power Apps CLI commands directly.

- **Deployment between tenancies**: An alternative to solutions deployment from the Power Pages solution is to use the CLI for deploying websites between different environments and different tenancies and also to migrate and convert website formats.

- **Automation and scripting**: Sarah can leverage the Power Apps CLI's scripting capabilities to automate repetitive tasks and streamline her development workflow. By writing scripts that utilize Power Apps CLI commands, she can automate tasks such as building, testing, and deploying Power Pages projects, saving time and reducing the risk of errors.

> **Tip**
> Further reading at *Microsoft Learn*: `https://learn.microsoft.com/en-us/power-pages/configure/power-platform-cli-tutorial`

Configuring the environment for seamless collaboration

To ensure a cohesive and collaborative development experience, Sarah needs to configure her environment effectively. Seamless Integration with Microsoft Power Platform Power Pages offers seamless integration with Dataverse, Power Automate, and the broader Power Platform, eliminating the need for any API work. This means you can effortlessly connect and utilize these Microsoft products directly within Power Pages, ensuring a streamlined and efficient development process. By leveraging built-in connectors and pre-configured integration points, you can focus on building your web apps without worrying about the technical complexities of integrating various services, thus enhancing productivity and reducing development time.

Version control integration

Sarah can integrate her development environment with VCSs such as GitHub. This enables efficient collaboration, change tracking, and project stability. Sarah can set up a GitHub repository dedicated to her project, providing a centralized location for storing and managing code. Project cloning within Visual Studio Code will allow her to work on the project locally and facilitate seamless collaboration with team members. Version control within Visual Studio Code enables tracking changes and revisions effectively. With GitHub, Sarah can regularly stage updates and commit modifications in Visual Studio Code for traceability, sharing updates by pushing changes to GitHub when ready. She also keeps her local copy synchronized by pulling updates from GitHub as needed. In case of conflicting changes, Visual Studio Code's built-in tools help Sarah resolve merge conflicts efficiently. Furthermore, she utilizes Visual Studio Code's capabilities to create and merge feature branches, allowing for organized development and streamlined integration of new features into the project.

Managing environments

Within the Power Platform admin center, Sarah can manage environments to create sandboxes for development and testing, promoting stability. This is where Sarah manages her environment storage and manual backups.

Security and access controls

Sarah configures access controls and security measures to ensure that development efforts align with Rob the Builder's security policies.

As Sarah embraces her role as a developer, she's equipped with the tools, knowledge, and collaborative environment needed to turn her creative visions into powerful digital solutions. The journey has only just begun, and as Sarah crafts the digital landscape for Rob the Builder, her expertise and the capabilities of Power Pages will join forces to drive a revolution in the company's operations. Sarah's journey continues as she delves into the realm of Power Pages design, leveraging the no-code/low-code prowess of Power Pages to shape the future of Rob the Builder.

Summary

The journey of Rob the Builder's transformation is one of inspiration and innovation. Sarah, armed with Power Pages and guided by Brenda's passion for modernization, embarks on a mission to revamp the company's back-office systems. The upcoming chapters will delve into the intricacies of Power Pages, from setup to development, illustrating how technology can be harnessed to reshape a business's destiny. With tools in hand and a clear vision, Rob the Builder's story demonstrates the potential of embracing change to drive growth, efficiency, and a brighter future.

The upcoming chapters will delve deeper into how Power Pages becomes the foundation for Rob the Builder's metamorphosis.

In the next chapter, we will look at setting up and examining the tools that Sarah needs to complete the website and app development.

2
Power Pages Design Studio

In the last chapter, we tackled the business challenges faced by Rob the Builder, highlighting the company's outdated website and systems that hindered business operations and project bids. Embarking on a transformative journey, Sarah and Brenda began exploring Power Pages, installing essential tools, and laying the groundwork for a digital overhaul of their business.

In this chapter, we will explore the exciting journey of Sarah and Brenda as they dive into the world of the Power Pages design studio to transform the digital presence of Rob the Builder.

In this chapter, we will cover the following topics:

- Planning the website structure
- Creating static pages
- Final touches and preparing for launch

Sarah was brimming with excitement for her meeting with Brenda. They planned to dive deep into developing new website content. This update was much needed, replacing their decades-old site. The original site was built years ago. It was basic, featuring a Home page, an **About Us** section, and listings of clients and projects. Brenda hadn't reviewed the site in years. Revisiting it with Sarah was both embarrassing and a relief. Sarah assured Brenda that the new site would be easy to update and maintain. She also promised to teach Brenda how to do it. This was a source of amazement and comfort for Brenda.

Planning the website structure

Moving forward from their reflections on the old site, Sarah and Brenda then shifted their focus to planning the structure of the new website. They sat down with the Power Pages design studio open in front of them, ready to turn their vision into reality. The first task at hand was reviewing the various templates provided by Power Pages, selecting one that best represented the essence of Rob the Builder. They knew their new site needed to be not only visually appealing but also functional and easy to navigate. As they browsed through the options, they discussed the key sections the website would feature:

- An **Home** page

- An **About Us** page

- A **Contact Us** page

- A client listing consisting of an image and text for each listing

Sarah decided to write an Agile user story to capture the essence of their website development work and ensure clear communication with the team and stakeholders.

User story – crafting a modern website for Rob the Builder

As a construction company aiming to enhance our digital presence and showcase our services effectively, we need a modern and user-friendly website. This website should reflect our brand identity, provide clear navigation, and engage visitors with compelling content and imagery so that we attract potential clients, communicate our expertise, and facilitate seamless interactions with our audience.

Acceptance criteria

Here are the acceptance criteria for building the website:

- The website should have a visually appealing design that aligns with our branding guidelines

- It should feature key sections such as **Home**, **About Us**, **Services**, and **Contact Us**, each providing relevant information to visitors

- The **Home** page should welcome visitors with an engaging headline, accompanied by high-quality images and introductory text

- The **About Us** page should provide insights into our company history, values, and team members, fostering trust and credibility

- The **Services** page should showcase our offerings in a clear and organized manner, detailing the range of services we provide

- The **Contact Us** page should include a contact form or clear instructions for visitors to reach out to us easily

- The website should be responsive and accessible, ensuring optimal viewing experience across devices

- Content should be organized logically, with intuitive navigation and consistent layout across pages

- The website should be hosted on a reliable platform and configured with proper security measures to safeguard user data

Tasks

Here is a list of tasks for building the website:

1. Collaborate with Brenda to finalize the website structure and content.
2. Select appropriate templates and design elements within the Power Pages design studio.
3. Populate pages with relevant text, images, and multimedia content.
4. Customize design elements, such as colors, fonts, and layout, to align with branding guidelines.
5. Test website functionality and responsiveness across different devices and browsers.
6. Iterate based on feedback from stakeholders and make necessary revisions.
7. Deploy the website to a production environment and configure domain settings for launch.

By implementing this user story, we aim to create a professional and engaging website that effectively represents Rob the Builder and attracts potential clients while providing valuable information to our audience.

Having written and submitted the user story to her client, Sarah would now go on to install Power Pages templates both to learn from and see if she could make use of their designs to repurpose them for her work.

Template selection and website vision

In the previous chapter, Sarah had already set up the environments and initial website using Power Pages. Microsoft's policy of offering a 30-day free trial and allowing users to create multiple websites within the same tenancy worked to the company's advantage. This flexibility enabled Sarah to experiment confidently with various template designs. She named these experimental sites `RobBuilderDev1`, `RobBuilderDev2`, and `RobBuilderDev3`. This approach allowed her to explore different design options without any immediate commitment, ensuring that the final choice for Rob the Builder's website would be well-informed and perfectly suited to their needs.

To create the websites, Sarah follows these steps:

1. Log in to Power Pages.
2. Select the **Dev** environment and select the **Templates** tab.
3. Scroll through templates, read the description, and view their styling design.
4. Select a **Preview** template such as **Starter layout 1**, as shown in *Figure 2.1*.
5. Having reviewed the description, select **Choose this template** to install it.
6. Enter a site name and a web address, such as `RobBuilderDev1`:

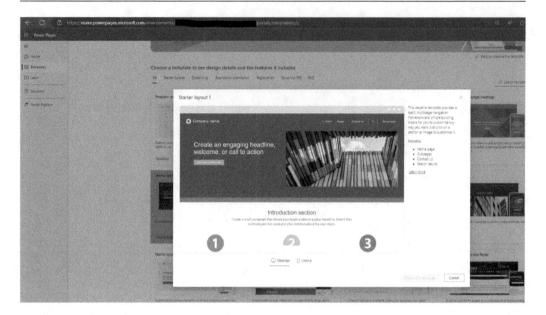

Figure 2.1 – Power Pages design studio templates

After thoroughly reviewing and selecting their preferred templates from Power Pages, Sarah now had a clear vision for the page layouts, font choices, and styling.

Developing the new website

They began the hands-on process of crafting the new website. Brenda opened a Word document and created headings for each web page, while they transferred and updated text from the old website. Brenda, deeply knowledgeable about the business, was decisive about the messaging. They spent several hours refining the content and wording. Over the years, Brenda had collected a treasure trove of photos and testimonial messages from satisfied clients. Retrieving these, they chose which images to use, finalized the text and bullet points, and highlighted customer comments to include. Brenda felt a growing sense of pride and accomplishment. Finally, she created a website that truly reflected their achievements and the essence of their business.

Building the home page

With a clear vision and all the necessary content at hand, Sarah and Brenda were ready to begin the actual construction of their new website in the Power Pages design studio. This phase was about turning their plans and designs into a digital reality, a step where Sarah's technical skills and Brenda's deep understanding of their business would come together.

> **Tip**
> For further reading on the Power Pages design studio, follow the link to *Microsoft Learn*: `https://learn.microsoft.com/en-us/training/modules/power-pages-studio/`.

Building the website pages

Sarah will begin with the home page. This section will provide a detailed, step-by-step guide on how Sarah uses the Power Pages design studio to build the home page, emphasizing how each element is added and customized. They had several sites listed as `RobBuilderDev1`, `RobBuilderDev2`, and `RobBuilderDev3` from the various templates they had configured. Now, they had a good idea of which template they wanted to use and the content they would implement for these first few pages. Sarah and Brenda decided to use the template they had named `RobBuilderDev1` to work on for their home page and to create a draft website on which to base their work.

Sarah followed these steps to access the Power Pages design studio:

1. Sarah opened the Power Pages design studio and edited `RobBuilderDev1`.

2. She selected the **Home** page from the **Pages** tab.

3. Sarah clicked on the text areas within the Home page designer to edit. Sarah pasted in the text that Brenda had previously prepared, including the welcoming headline and introductory text.

4. Sarah uploaded the images that Brenda had prepared. In some cases, she used an image editor to crop or resize images to fit the image controls placed in the template. The preselected template already had images, so Sarah replaced these by clicking on them and uploading the new images, as shown in *Figure 2.2*:

Figure 2.2 – Replacing and uploading images

5. As they developed the **Home** page together, they discussed some of the text box and image design and made a few changes to the original template design that they agreed would be better; for example, making some headlines a bit smaller:

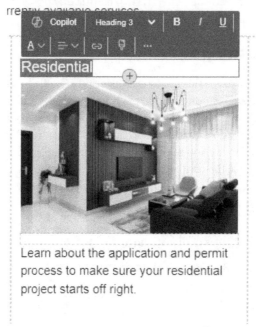

Figure 2.3 – Clicking the text brings up the style tool

6. Sarah used the design tools on the page to customize colors, fonts, and other style elements to match the company's branding guidelines. Now, they added branding elements such as the company logo, tagline, or any other unique branding features. There was already a logo in the header from the template. Sarah clicked on it and replaced it with the company logo Brenda had prepared:

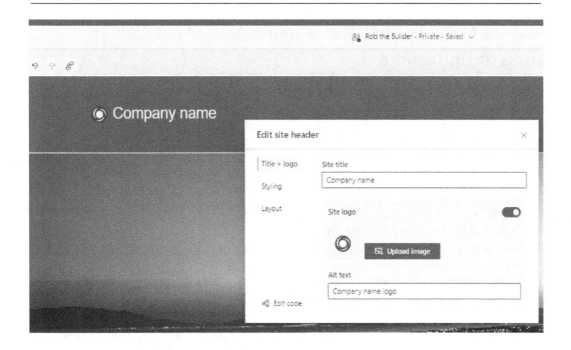

Figure 2.4 – Clicking the logo to replace it

7. When they were satisfied with the page content, Sarah used the **Preview** feature in the Power Pages design studio to see how the **Home** page looked in a live environment.

8. Having built the **Home** page, Sarah and Brenda spent the rest of the day applying this process to other pages. They focused on the **About Us** and **Contact Us** pages.

9. For the client listing page, they started with a new page and selected sections to insert, selecting two-column sections, creating a listing of headings, followed by an image and then a block of text, highlighting keywords with bold and colorful text.

10. They then worked on the footer. Sarah had some basic footer text that she had authored herself, and Brenda had some ideas; they also placed and inserted links for their social media pages.

Now that Brenda and Sarah had created a home page along with the pages to showcase their business, they were ready to replace their legacy website.

Brenda now wanted to discuss with Sarah what the next steps would be. She wanted a website that not only interacted with her customers and suppliers but also with the construction teams or, more specifically, the foremen to implement the **incident management (IM)** pages.

Content management and organization in the Power Pages design studio

Effective content management is crucial for a successful website. In the Power Pages design studio, organizing content begins with a clear understanding of the site's purpose and audience. Sarah and Brenda used this platform's tools and features to make content organization both intuitive and efficient. We will cover the following in this section:

1. **Defining the website's hierarchy**: Identifying main pages such as **Home**, **About Us**, **Services**, and **Contact Us** and creating sections for easy navigation.

2. **Creating and organizing pages**: Determining specific pages needed under each main section and adding them using the Power Pages design studio.

3. **Maintaining a consistent layout across pages**: Maintaining a uniform look and feel by using similar templates or layouts.

4. **Pointing a custom domain to a new website**: Steps to redirect the company's domain from the legacy website to the new one created in the Power Pages design studio.

This section will highlight Sarah's and Brenda's collaborative efforts in organizing content effectively and seamlessly transitioning to their new website.

Defining the website's hierarchy

Sarah started by outlining the hierarchy of the website. This step involved identifying main pages such as **Home**, **About Us**, **Services**, and **Contact Us**. The Power Pages design studio's user-friendly interface allowed her to create and name these sections easily, forming the foundation of the site's structure.

Creating and organizing pages

Under each main section, Sarah determined the specific pages needed. For example, under **Services**, different pages were created for each service offered. She utilized the Power Pages design studio to create these pages, organizing them under the appropriate sections to ensure a logical flow of information. Sarah added new pages by selecting + **Page** in the studio, as shown in *Figure 2.5*.

Here, Sarah added the page name and its URL path:

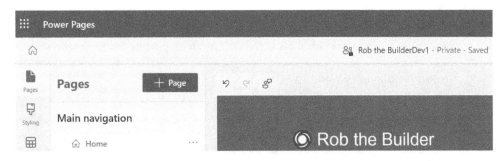

Figure 2.5 – Adding a new page

Brenda and Sarah discussed additional pages, such as a **Finance** section with **Invoices**, **Orders**, and **Quotes** subpages. They also discussed the **Services** section, considering adding **Projects** and **Incident Management** subpages. Using the Power Pages design studio, Sarah was able to add subpages by selecting **Add a new subpage**, as shown in *Figure 2.6*:

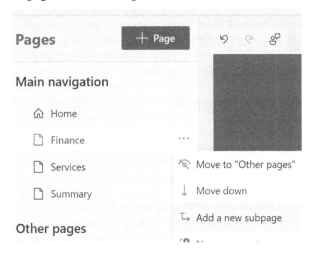

Figure 2.6 – Adding a subpage

Brenda mentioned the need for a **Finance** page. It would include **Invoices**, **Orders**, and **Quotes** subpages. They also discussed what would be under **Services**: maybe to include **Projects** and change **Order requests** and **Incidents**.

Sarah simply created new pages and subpages in the Power Pages design studio and dragged one under another until they had navigation in an order that Brenda thought looked good. Sarah remembered how her tutor at college had said that you could start a website without **wireframe modeling** using a graphics-type tool and how hard and awkward they were to use. She thought to herself, *Wireframing a project is easier directly in Power Pages!*

> **What is wireframe modeling?**
>
> Here is a link to get more information on wireframe modeling: `https://support.microsoft.com/en-us/office/use-wireframe-templates-to-design-websites-and-mobile-apps-2d54dc55-f5c4-49a2-85da-d649eb7fc281`

Maintaining a consistent layout across pages

To maintain a consistent look and feel across the site, Sarah focused on using similar templates or layouts for similar types of pages. This approach, easily managed in the Power Pages design studio, helped to keep the navigation intuitive and the user experience cohesive.

Sarah started with the default web template that is used on every page, but she had in mind that eventually, she would need to create her own web template later, especially for pages such as **Incidents**.

The process of content management and organization was a collaborative effort between Sarah and Brenda. As they worked together in the Power Pages design studio, they found that the platform's capabilities greatly simplified what could have been a complex task. Their website's content was now well organized, easily navigable, and poised to effectively communicate with their audience.

Pointing a custom domain to a new website

Brenda now wanted to point the company's own domain from the legacy website to the new website. Here are the steps that Sarah followed:

1. **Access domain registrar account**: Sarah logged in to the account where her current domain was registered (for example, GoDaddy, Namecheap) and navigated to the section to manage her domain.

2. **Locate DNS settings**: Within her domain registrar's dashboard, she found the domain that she wanted to point to on the new website. From here, she accessed the DNS settings/management area.

3. **Update DNS records**: Sarah then navigated to the page where she was able to edit her **A records** and **CNAME records**. This is where she was able to set directions for where to point the domain.

   Here, Sarah changed the A record to point to the IP address of her new website hosted on Power Pages. Alternatively, if Sarah was using a CNAME record, she would similarly change it to the appropriate text provided by Power Pages.

4. **Find your Power Pages site's IP address or URL**: In the Power Pages design studio, Sarah located the section that provided details about her site's hosting information. Here, she copied the IP address that the Power Pages site was hosted on (if she was using CNAME, she would have copied the URL).

5. **Input the IP address or URL in the DNS record**: Sarah went back to her domain registrar's DNS settings and pasted the IP address into the A record field (if using CNAME, she would have pasted the URL).

6. **Save DNS settings**: After updating her records, she saved her changes.

   Keep in mind that DNS changes can take up to 48 hours to fully propagate across the internet.

7. **Verify domain connection in the Power Pages design studio**: Once the DNS changes had been made, Sarah went back to her Power Pages design studio. Here, she navigated to the **Domain Management** section and verified that the custom domain was pointing to her new Power Pages site correctly.

8. **SSL certificate**: Navigating to the website, Sarah confirmed that her web page was secure, to ensure the site was safe to use. She did this by looking at the padlock button to the left of the URL. Power Pages handled this for Sarah.

9. **Test the website**: Further, Sarah ensured that the website loaded correctly and that all links and functionality were working as expected.

Tip

It's important to ensure that you have a backup of any existing data or email services tied to your domain before making changes to the DNS settings to avoid any unintended disruptions.

For further reading on adding a custom domain name, follow the link to *Microsoft Learn*: `https://learn.microsoft.com/en-us/power-pages/admin/add-custom-domain`

By following these steps, Sarah successfully pointed the existing domain to the new website created in the Power Pages design studio, ensuring a seamless transition for users and maintaining the company's brand presence online.

Developing websites and deploying them to production

Having established a production website and several development sites, Sarah faced a new challenge. She needed to deploy content from one environment to another. She had built the website with Brenda in the Rob the Builder development environment that she had created in *Chapter 1*, now she needed to deploy it to the production environment she had created in *Chapter 1*. To do this, she would use the Power Pages solution tool to add her website and then deploy the solution to the production environment. Here are the steps for the same:

1. In the studio, Sarah selected the **Home | Solutions** tab.

2. She then clicked **New solution** and entered a name, as shown in *Figure 2.7*.

3. Within the solution, Sarah added the website and pressed **Export**.

4. Once that was complete, Sarah downloaded the `.zip` export to her local PC.

5. In the studio, Sarah selected the production environment and pressed **Import**. After that, she chose the downloaded `.zip` file to import:

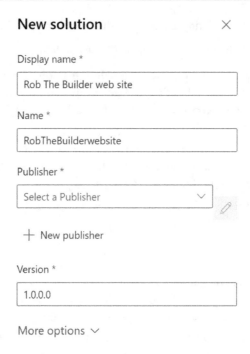

Figure 2.7 – New solution for deploying the website

Now that the website was deployed, it was still set as private. It was time to set it to **Public** so that other users and the public could view the new website.

Setting the website to public

Sarah selected the settings to set the website as public, and they tested it on their phones to see how it looked.

To do this, in the studio production environment, Sarah selected the **Set up** tab and set **Site visibility**, as shown in *Figure 2.8*, to **Public**:

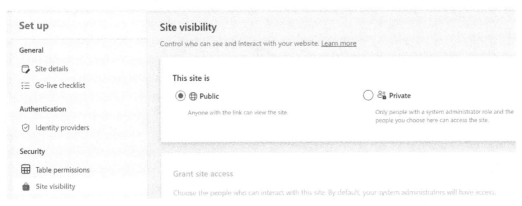

Figure 2.8 – Site visibility

The website static pages were ready, and Brenda was ready to show them to her colleagues and announce them to her customers.

Summary

In this chapter, we followed Sarah's and Brenda's journey as they delved into the Power Pages design studio to revamp Rob the Builder's digital presence. They started by planning the website's structure, selecting templates, and drafting a user story to guide development. With the Power Pages design studio's flexible trial period, they experimented with various designs before settling on a vision for their site.

Once they established their template and content, Sarah led the development process. Using the Power Pages design studio, they created **Home**, **About Us**, and **Contact Us** pages, meticulously customizing each element to align with their branding. They also explored the process of building a client listing page, adding images, and refining text.

Sarah and Brenda then tackled content management and organization, defining the website's hierarchy and ensuring a consistent layout across their web pages. They effortlessly added new pages and subpages using the Power Pages design studio's intuitive interface, streamlining their website's structure.

To finalize the transition, Sarah guided Brenda through pointing the company's domain to the new website, ensuring a seamless experience for users. With the website deployed to the production environment, they set it to public, ready for viewing.

Having successfully created the static pages of the website and deployed them to the production site, Sarah's next goal was to enhance its visual appeal. She aimed to use the Power Pages design studio's styling tools to develop a distinct style that resonated with the company's branding. This would lay the groundwork for the more intricate pages they had planned, as discussed in their navigation strategy.

In the next chapter, we'll explore how Sarah utilized the styling tools in the Power Pages design studio to achieve this, taking a deep dive into creating a cohesive and visually appealing brand identity for Rob the Builder's website.

3

Power Pages Studio – Styling and Themes

In the previous chapter, we ventured into Power Pages Studio, where Sarah, the developer, worked alongside her client, Brenda, to create and publish the static pages of their website. As this chapter unfolds, we'll dive into the next phase, which is leveraging Power Pages Studio's styling and theme tools to bring a new brand identity to life for the website. Following the successful launch of their updated site, Brenda brought news from Rob, the CEO: a graphic design firm was commissioned to refresh the company's branding. The firm delivered a "branding book," complete with specific fonts, primary and secondary color schemes, and styling guidelines for various web elements, such as buttons and links.

Brenda convened this meeting to share these new branding guidelines, indicating the need to revise the website's current styling so that it aligns with this updated brand image. In prior discussions, Brenda had also mentioned the necessity for additional pages to cater to client and supplier interactions. Sarah recognized the magnitude of this task. The method of applying styling piece by piece, as done previously using Studio tools, would not suffice for the comprehensive updates required. Anticipating the need for a more scalable solution, she contemplated working on the site's themes and creating a new **Cascading Style Sheets** (**CSS**) file for more efficient management of the site's styling. Prepared with extensive research on this technique, Sarah was ready to tackle the challenge. Upon receiving the branding guidelines from Brenda, she committed to adapting the website to these new standards and keeping Brenda informed of her progress.

In this chapter, we will cover the following topics:

- Accessing CSS features in Power Pages Studio
- CSS techniques
- Applying a custom theme with CSS
- Implementing the branding guidelines

This chapter aims to guide Sarah through the intricate process of aligning the website with the new branding guidelines, employing advanced styling techniques, and ensuring that the final product resonates with *Rob The Builder's* newly established brand identity.

As Sarah prepared to embark on the significant task of rebranding the website, she knew that a deep understanding of CSS was crucial. CSS is the tool that would enable her to transform the new branding guidelines into a digital reality.

> Tip
> CSS is a language that's used to style the appearance of a website, dictating everything from layout to font styles.

Introduction to CSS

CSS is more than just a set of rules for styling a website; it's the cornerstone of modern web design. It allows developers such as Sarah to separate content from design, ensuring that web pages are not only functional but also aesthetically pleasing.

CSS plays a vital role in implementing a website's branding. It offers the flexibility to apply specific color schemes, typefaces, and other stylistic elements consistently across multiple web pages. This consistency is key to establishing and maintaining the brand identity throughout the website.

One of CSS's greatest strengths is its ability to control the layout of multiple web pages from a single file. This means Sarah can make widespread changes to the site's appearance without having to edit each page individually. It's an efficient way to apply the new branding guidelines provided by Brenda.

Moreover, CSS is crucial for responsive design. With an increasing number of users accessing websites via mobile devices, CSS allows Sarah to ensure the site looks good and functions well on all screen sizes. She can define different styling rules for different screen widths, making the website adaptive and user-friendly.

Understanding CSS will empower Sarah to fully realize Brenda's vision for the website. It's the tool that will bring the new brand identity to life in the digital space, ensuring that every visitor to *Rob The Builder's* website experiences the essence of the brand just as intended.

> Tip
> Further reading on CSS can be found at `https://www.w3schools.com/css/css_intro.asp`.

Modifiable styles in Power Pages Studio

In Power Pages Studio, Sarah can modify a variety of styles to align with her branding guidelines. These include the following:

- **Typography**: Fonts, font sizes, font weights, line heights, and letter spacing
- **Colors**: Primary and secondary color schemes, background colors, text colors, and border colors
- **Layout**: Margins, padding, alignment, and positioning of elements
- **Buttons**: Button styles, sizes, colors, hover effects, and border styles
- **Links**: Link colors, hover effects, and text decorations
- **Forms**: Input field styles, labels, and form layout
- **Navigation**: Menu styles, drop-down styles, and navigation bar layout
- **Headers and footers**: Styles for headers and footers, including background colors, text styles, and alignment
- **Images and media**: Styles for image sizes, border styles, and alignment

With the task of implementing the new branding guidelines at hand, Sarah's first step was to access the CSS features within Power Pages Studio. This toolset allows developers to directly manipulate the styling of a website with precision and flexibility.

Accessing CSS features in Power Pages Studio

To access these CSS features, Sarah navigated to the main dashboard of Power Pages Studio and selected the **Styling** tab. This tab provides access to the themes offered by Power Pages for every website. Sarah has chosen a theme that was closest to the branding. For example, there were themes with dark and white backgrounds and color schemes that looked good. Having selected a theme, Sarah was then able to change many of the common styles for the body color, fonts, and objects such as buttons and links. She made changes, selecting the colors from the branding book, and it started to look close to what she needed. Sarah was able to save and preview the existing pages to check how the style changes appeared. Here is a screenshot of the **Styling** tab:

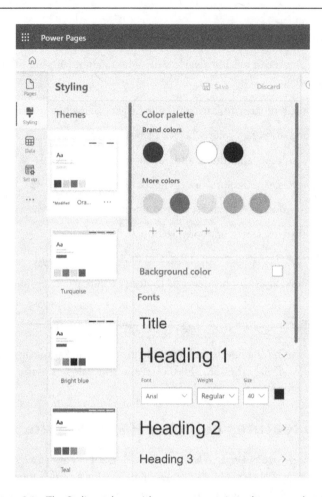

Figure 3.1 – The Styling tab provides access to various themes and styles

Power Pages Studio's interface made it easy for Sarah to interact with CSS. The platform provides a user-friendly codeless CSS editor that allows for rapid styling changes to be made and applied across the theme.

The CSS editor in Power Pages Studio also allows for real-time previews, meaning Sarah could instantly see the effects of her changes. This feedback was useful for her to experiment with different styles and ensure that everything looked perfect before she applied the changes to the live site.

Accessing the CSS code

Sarah can access the CSS code for the theme, which allows her to review the code based on the modifications she has implemented. This approach serves as a practical learning method as she can make changes using the styling tool and then examine the corresponding alterations in the CSS code. To access the theme CSS code in the Visual Studio Code editor, as shown in *Figure 3.2*, Sarah followed these steps:

1. In Power Pages Studio, Sarah selected the **Edit Code** link.

2. Sarah was taken to the Visual Studio Code editor. From here, she browsed to the **Web Files** link.

3. Finally, Sarah selected the `themes.css` file to display the CSS code, as shown in *Figure 3.2*:

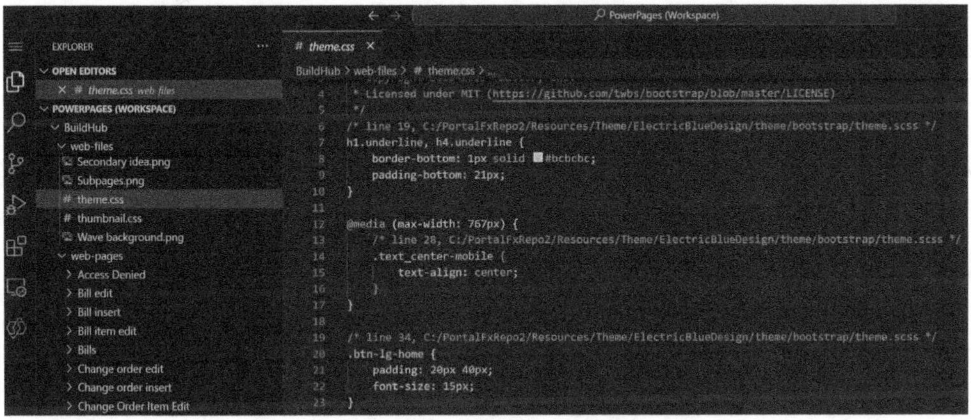

Figure 3.2 – CSS code for themes

Despite this, Sarah was eager to deepen her understanding of custom CSS. She dedicated time to research and learning how to craft her own CSS files:

Figure 3.3 – Opening the Themes CSS

Rob the Builder was an accommodating client, and Sarah knew she had the opportunity to refine her skills. She was determined to gain the proficiency needed to create custom CSS files and develop unique themes. Sarah realized that mastering this skill was essential for her to evolve into a fully fledged professional in Power Pages development. This expertise would not only benefit her current project with *Rob the Builder* but also lay a solid foundation for her future endeavors with other clients.

Implementing basic CSS customizations

Sarah's initial modifications in the Studio theme tool (*Figure 3.1*) laid the foundation for basic CSS customizations. She updated the site to reflect the brand's color schemes and adjusted the sizing of headings, buttons, and link colors. She also tweaked the layout, modifying margins and alignments for a cleaner presentation.

Create a custom CSS file

After laying the groundwork with basic styling, Sarah was ready to take the next step: creating a CSS file for more customized work. In the Studio, within the theme where she had made changes, Sarah copied the existing code. She then created a new file, pasted the copied code into it, and named it `RobBuilderCustomCode.css`. Following this, she proceeded to upload it to the site.

To create a custom CSS file that will append the CSS with new styles and override existing styles, Sarah did the following:

1. First, she created a file with a `.css` extension.

2. Then, she opened Power Pages Studio and selected the **Styling** button.

3. After, Sarah selected a theme and the ellipsis (**...**) button.

4. Next, she selected the **Manage CSS** option, which opened the **Manage CSS** section, as shown in *Figure 3.3*.

5. Once she'd done this, Sarah uploaded the new CSS file – that is, `RobBuilderCustomCode. css`.

6. Sarah could now edit this file in Visual Studio Code under **Web Files**:

Figure 3.4 – Uploading a new CSS file

By uploading a new CSS file, Sarah can apply her custom styles to the entire website, allowing her to precisely control the appearance of various elements and ensure consistency with the new branding guidelines.

> **Tip**
>
> Further reading on adding a custom CSS file can be found at `https://learn.microsoft.com/bs-latn-ba/power-pages/getting-started/tutorial-add-custom-style`.

With the basics in place, it was time for Sarah to learn more advanced CSS techniques. We'll explore these in the next section, starting with working with custom fonts.

Custom fonts

To learn how to implement custom fonts, Sarah looked at implementing Google Fonts.

Google Fonts

Google Fonts offers a wide variety of fonts that can be easily integrated into a website. They are useful because they provide a simple way to enhance the visual appeal of a website without the need for complex font management. Google Fonts gives Sarah easy access to a larger library of custom fonts.

To implement custom Google Fonts, Sarah referenced Google Fonts in a content snippet or added them to the web template. She then included the font in the CSS using the following syntax:

```
font-family: 'font name';
```

Sarah reviewed the guidance notes on the corporate branding, as agreed with Brenda. However, the fonts specified in the guidance were not on the list in the Themes style editor. Sarah needed to add custom fonts to the website that were in keeping with corporate branding style guidance. So, she researched how to add new fonts.

Importing custom fonts

Despite the ease of using Google Fonts, Sarah still needed to upload a custom font that was not available on Google Fonts. This required her to download a font file and reference it within her CSS code. Here are the steps she followed to import custom fonts:

1. First, Sarah downloaded the font as a file to her PC.
2. Then, she created a new web file through Power Pages Management.

3. Next, Sarah attached the Hero font file to the web file record.

4. Once she'd done this, Sarah amended the web file's partial URL so that it could be referenced as /Hero, as shown here:

```
@font-face {
    font-family: '< font name>';
    src: url('../< font name.ttf>');
}
```

5. After opening the CSS code (*Figure 3.2*) and referencing the new fonts, Sarah adjusted the styles in the CSS:

```
@font-face {
    font-family: 'Hero';
    src: url("/herofont.ttf");
}
@font-face {
    font-family: 'Nunito Sans';
    src: url("/nunitosans.ttf");
}
h1, h2, h3 {
    font-family: 'Hero', sans-serif !important;
}
h4, h5, p {
    font-family: 'Nunito Sans', sans-serif !important;
}
```

The result was satisfying – the new fonts seamlessly integrated into the static pages, all consistently styled through the theme.

Changing the website's logo

Next, Sarah needed to change the website's logo. To do so, she did the following:

1. First, Sarah opened Power Pages Studio and selected the home page.

2. Then, she selected the header.

3. Here, Sarah found the **Edit site header** link. Upon clicking it, she was provided with the form shown in *Figure 3.4*.

4. Finally, Sarah uploaded a site logo image:

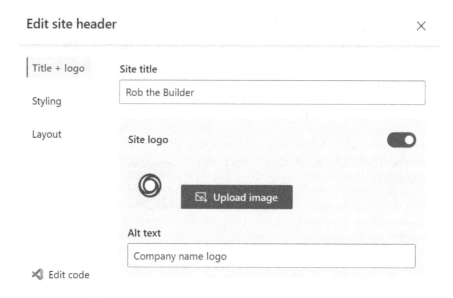

Figure 3.5 – Editing the logo

By doing this, Sarah replaced the image logo and learned how to add more images to the site.

Adding web file images

Brenda had given Sarah three high-quality images to incorporate from their past projects. Sarah added the images to the website as Web Files by doing the following:

1. From Power Pages Studio, Sarah clicked the ellipsis (**…**) and opened **Power Pages Management**.

2. Then, she browsed the **Content** tab and opened the **Web Files** tab, as shown in *Figure 3.5*.

3. To create a new web file, Sarah selected **New**.

4. From here, Sarah entered the name of the file, a partial URL, and selected the website – that is, **Rob the builder 1**.

5. Sarah saved her amendments and then attached the file under **File Content**. She chose the file that Brenda had given her.

6. Sarah repeated this for each new image she needed to add to the website:

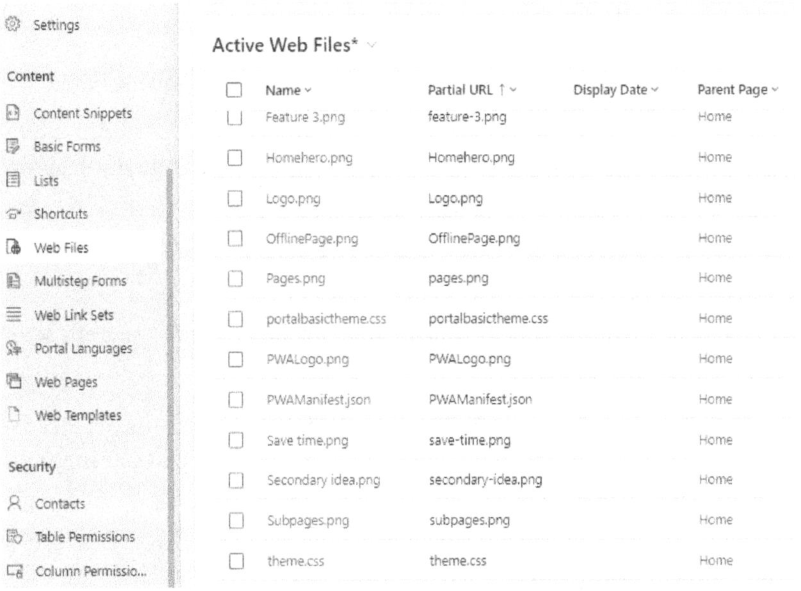

Figure 3.6 – Web Files in Power Pages Management

These new images were now available to be used on the website and could be referenced via their partial URLs. She asked her client where these images needed to go. This led her to having to learn more advanced styling techniques.

Incorporating advanced CSS techniques

Rob the CEO told Brenda that he was pleased with the home page but he now wanted a carousel of the new images on the home page displaying images of their past projects in a loop.

Sarah researched how to add a carousel and came up with the following CSS:

```
.carousel {
  display: flex;  overflow-x: auto;  scroll-snap-type: x mandatory;
}
.carousel-item {
  flex: none;  scroll-snap-align: start;  width: 100%;  position:
relative;
}
.carousel-img {
  width: 100%;  display: block;  border-radius: 10px;
}
.carousel {
  scroll-behavior: smooth;
}
```

Let's take a closer look at this code:

- `.carousel`: This class makes the carousel a flexible container, allowing horizontal scrolling with mandatory snap points
- `.carousel-item`: Each item in the carousel is positioned horizontally, snaps to the start, and takes up the full width of the carousel
- `.carousel-img`: The images within each item fill the container's width and have rounded corners for a polished look
- `scroll-behavior: smooth`: This optional property enables smooth scrolling for a better user experience

Having completed the CSS for the carousel, Sarah researched the HTML to use to make the carousel work. She pasted this HTML as the top of the home page's code in a page-wide section so that the carousel appeared under the header:

```
<div class="carousel">
  <div class="carousel-item">
    <img src="/carousel1.jpg" alt="Project Image 1" class="carousel-
img">
  </div>
  <div class="carousel-item">
    <img src="/carousel2.jpg" alt="Project Image 2" class="carousel-
img">
  </div>
  <div class="carousel-item">
    <img src="/carousel3.jpg" alt="Project Image 3" class="carousel-
img">
  </div>
</div>
```

In this example, the CSS creates a simple, scrollable carousel.

Each `carousel-item` represents an individual slide in the carousel. The images (`carousel-img`) are set to fill the width of their container for a responsive design. This structure provides a basic yet functional image carousel for the home page, showcasing *Rob the Builder's* past projects. Sarah can further customize the carousel's appearance and behavior by modifying the CSS and HTML as needed.

Summarizing Sarah's journey through this chapter, we highlighted the key aspects of her work in enhancing the website's styling and themes.

Summary

In this chapter, we journeyed through the advanced aspects of Power Pages Studio, focusing on styling and themes to enhance a website's brand identity.

Sarah accessed and utilized the various CSS features in Power Pages Studio, selecting themes closest to the new branding guidelines and customizing common styles such as body color, fonts, and button links. She tackled the challenge of incorporating custom fonts that weren't initially available in the theme style editor. Sarah successfully added them as Web Files and integrated them into the CSS, reflecting the new branding's typography across the site.

Stepping up her CSS game, Sarah added a carousel to the home page, showcasing *Rob the Builder's* past projects. She learned and implemented advanced CSS techniques for a more dynamic and interactive website.

In this chapter, we've seen Sarah's growth as a developer, mastering CSS customizations and utilizing advanced techniques to bring a refreshed, modern look to the website. Her dedication and adaptability in learning and applying new skills underscore the evolving nature of web development.

In the next chapter, we will delve into creating an incident management feature to demonstrate the **Create, Read, Update, Delete (CRUD)** pattern in Power Pages. This chapter will focus on efficiently planning and building this feature while showcasing practical applications of Power Pages Studio in real-world scenarios.

4

Dataverse Tables and Forms

In the previous chapter, we delved into the world of Power Pages Design Studio, where Sarah and Brenda, guided by the newly established branding guidelines, transformed Rob the Builder's website. Continuing their journey, in this chapter, they face a new challenge: developing an incident management system for Rob's construction business. Brenda emphasizes the importance of this system in tracking accidents and incidents, stressing the need for a formal report adhering to the **Occupational Safety and Health Administration** (**OSHA**) standards. Sarah prepares to tackle this essential feature by implementing the **CRUD pattern** (short for **create, read, update, delete**), which is a common pattern for Power Pages development.

Sarah will be leveraging Dataverse, an integral component of Power Pages, to manage data effectively in her incident management system. Dataverse provides a structured and organized environment for creating tables, defining relationships between them, and managing forms. By utilizing Dataverse, Sarah can ensure data integrity and facilitate seamless interaction between different components of her system. Moreover, Dataverse's architecture complements Power Pages by providing a robust foundation for storing and accessing data, enhancing the scalability and reliability of the entire application.

One prominent feature of Dataverse is its forms. By utilizing Dataverse forms within Power Pages, Sarah can seamlessly create, customize, and embed forms into her web application. These forms are directly linked to the tables and data stored in Dataverse, allowing for easy data entry, editing, and viewing. Furthermore, this enables Sarah to build applications that leverage the structured data management capabilities of Dataverse.

Sarah embarks on the task of planning and constructing an incident management system, a vital component for the efficient operation of Rob the Builder's business. This chapter serves as a practical guide to developing and extending such features. You will learn how Sarah utilizes Power Pages Design Studio and Dataverse solutions to implement tables and forms. You will also understand the step-by-step process of translating a conceptual design into a functioning system.

We'll cover the following topics in this chapter:

- Planning the feature
- Developing a data model for tables and relationships

- Creating and configuring fields
- Creating and configuring Dataverse views to be used on web pages
- Creating and configuring Dataverse forms

Planning your development

As Sarah and Brenda sit down to discuss the development of the incident management pages, Brenda reminds Sarah of the critical nature of this project. They agree on the importance of meticulous planning to ensure a smooth development process, minimizing the need for time-consuming revisions later on. They consider employing an Agile approach, which is perfect for their dynamic, evolving project. Sarah notes that the key elements of the CRUD pattern for this project will include a list page and an edit page, each with action buttons, essential for managing incidents efficiently.

Planning and specifying the tables and field types correctly is crucial because rolling back changes to tables after implementing forms and Power Pages can be a time-consuming and frustrating process. If you are working in a group, you could use Agile to help you govern your work practices. This is a good, iterative process that involves planning, developing, reviewing, testing, and refactoring.

The key elements of the CRUD pattern for the incident management feature include a **list page** and an edit page with action buttons. The **list page** is a read type of landing page that lists records with action buttons for creating and inserting new records, an edit button to update records, and a read-only form page for closed records, such as closed or canceled incidents.

Developing the user story

To define the work requirements of her customer, Sarah decides to write an Agile story to plan the feature.

"As a construction company manager, I want to track incidents and accidents on the job site efficiently so that I can ensure the safety of my workers and comply with regulatory standards."

Acceptance criteria:

- The user should be able to log in to the incident management system securely
- The user should be able to view a list of all recorded incidents, including details such as date, type, severity, and status
- The user should be able to add new incidents to the system, providing details such as date, time, location, description, and type of incident
- The user should be able to edit existing incidents to update their details or status
- The user should be able to mark incidents as resolved or closed once they have been addressed
- The system should provide alerts or notifications for high-priority incidents or overdue tasks

- The user should be able to generate reports or summaries of incidents for analysis or compliance purposes

- The system should be accessible and user-friendly on both desktop and mobile devices

Planning the data model

Sarah begins by mapping out a detailed data model for the incident management system. She meticulously lists the necessary tables, fields, and their types, fully aware that correct initial decisions are crucial for a smooth development path. She emphasizes the importance of choosing the right field types and date formats from the onset, knowing well the challenges of altering these elements later. Sarah and Brenda discuss the need for multi-line text fields for comprehensive incident descriptions and the importance of precise data formats in incident reporting.

Note that Sarah can't change field types, so she must get these types right; otherwise, she will need to delete badly designed fields and recreate a field with its correct type. Also, note that once Sarah selects a date format, it can't be reversed, so she must decide on her date format. In our example, we're using the default local date.

Important note

You can learn more about different data types at `https://learn.microsoft.com/en-us/power-apps/maker/data-platform/types-of-fields`.

In the next section, we will explore the relationships between tables, providing an understanding of the data model's structure.

List and reference data

Sarah carefully considers the structure and relationships within her data model. She explains to Brenda the significance of list and reference data in their system. They discuss the implications of using choice fields versus separate reference data tables. Sarah opts for the latter, valuing their flexibility for future changes and ease of filtering on web pages. She knows that this decision will greatly influence the adaptability and functionality of their incident management system.

Sarah needs to be careful in her design choices. A choice field (for example, a list) is easiest to implement and use, but she must be mindful of its limitations. Once the page is released and has gone live, Sarah can't retire or refactor this choice field as it affects live data if a choice value has been used. It is more awkward to filter this choice field on a web page.

To ensure future adaptability and effective filtering, it's advisable to create a separate reference data table and establish a many-to-one (N:1) relationship between her main table (for example, the incident table) and the reference table (for example, the incident type). This approach proves beneficial when the reference data is expected to change or when the list must be filtered on a web page.

When listing tables, also list the relationships between tables.

> **Important note**
>
> Tables can have a field that is a lookup to another table; this is a many-to-one (N:1) relationship. Tables can also have many-to-many (N:N) relationships – for example, many contacts can be related to many tables.
>
> You can learn more about table relationships at `https://learn.microsoft.com/en-us/power-apps/maker/data-platform/create-edit-entity-relationships#types-of-table-relationships`.

With Sarah's data model complete, she must plan and create her forms using the Dataverse form editor.

Planning out the Dataverse forms

Next, Sarah outlines the structure of the Dataverse forms, aiming for an interface that's both organized and intuitive. She collaborates with Brenda, deciding on the layout and number of tabs and columns for each form section. They agree on a single tab layout for simplicity, envisioning a user-friendly interface that will work seamlessly on both desktop and mobile platforms. This step is pivotal in ensuring that the forms meet both their functional needs and the ease-of-use standards for Rob the Builder's team.

Power Pages produces a mobile version of her Power Pages forms. She will create a new Dataverse form and a tab for each form used. She can reuse a form tab for different Power Pages forms. For example, Sarah will use the same form for the read-only and edit pages of the incident management pages.

Mastering this CRUD pattern not only enables Sarah to develop wireframes and proof-of-concept pages quickly but also provides a more effective means of demonstration. Additionally, it serves as a foundation for the actual project since much of the work can be utilized in the development process.

After finalizing the data model and completing the Dataverse forms, Sarah must now identify the web roles that are used in the feature and configure table permissions to ensure the security of the feature.

Planning out the web roles

As they dive deeper into the planning phase, Sarah and Brenda discuss the various user roles that will interact with their incident management system. They understand that defining these roles is critical for establishing appropriate access levels and ensuring data security. Brenda suggests having distinct roles such as "Back Office" and "Foreman," each with specific access permissions, aligning with the operational structure of Rob's construction business. Sarah notes this down, preparing to define these roles precisely in the Power Pages setup. These roles act as personas that define access levels to data and web pages. Each role will have distinct responsibilities within the application. In the next section, we will focus on defining precise permissions for each web role to ensure data security and controlled access.

Planning table permissions configuration

Securing the application is crucial for protecting sensitive data and ensuring proper access controls. In this section, we will focus on planning the table permissions configuration. By carefully defining the permissions that are required for each web role, Sarah can establish granular control over data operations such as create, read, write, delete, and append/append to. Taking the time to plan this configuration will contribute to a robust and well-protected system.

Sarah lists the tables and permissions for each web role. Here, Sarah writes out a horizontal table showing the table permissions for the Foreman web role:

Table Name	Read	Write	Create	Delete	Append	Append To
Incident	Yes	Yes	Yes	Yes	Yes	Yes
Incident type	Yes	No	No	No	No	Yes
Injury type	Yes	No	No	No	No	Yes

Table 4.1 – The Foreman web role's table permissions

The preceding table shows the table permissions that must be configured for each table in use in the incident management feature for the Foreman web role.

> **A tip for understanding Append and Append To**
>
> The "Append To" privilege on the table (parent) is referred to in the lookup so that it can set the values for the lookups of this table on any other form. An example of this is "Incident" (child), which has a lookup of "Incident Type" (parent). So, here, the web role needs to have the "Append To" privilege on the "Incident Type" table to be able to set the incident type for the incident. Therefore, the web role needs to have the "Append" privilege on the "Incident" table to be able to set the "Incident Type" lookup.

Once Sarah has planned the table permissions and web roles for the security model, she can plan how best to implement the database design, something she will implement in the next chapter.

Planning your implementation in the recommended order

To ensure an efficient and seamless development process, it is recommended to follow a specific order of implementation. With this order, Sarah can systematically complete the required components and have them available when needed. For example, Dataverse forms must be created before basic Power Pages forms can be created. In this section, Sarah outlines the order for implementing the components:

1. Create tables, fields, and views in Dataverse for incidents and incident types.
2. Develop Dataverse forms for inserting, editing, and viewing incidents.

3. Configure table permissions for all relevant tables.

4. Build basic forms for inserting, editing, and viewing incidents.

5. Set up a list for displaying incidents.

6. Create web pages for the list, as well as inserting, editing, and viewing incidents.

7. Enhance the list by adding action buttons for quick insert and edit actions.

8. Assign roles to page permissions, ensuring proper access controls.

To access Dataverse's various tools, Sarah opens Power Pages Design Studio, as shown in *Figure 4.1*. By clicking on the **Data** tab, Sarah can manage tables, views, and Dataverse forms:

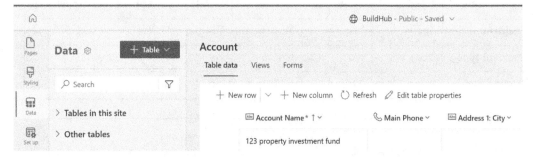

Figure 4.1 – Accessing Dataverse in Power Pages Design Studio

Now that Sarah has planned her work, in the next section, we'll learn about developing and implementing the data model and why the decisions Sarah makes are important to provide a good user experience when users interact with the incident management feature.

Practical example – developing the data model

In this section, we will walk through a practical example of developing the data model for an incident management feature. By examining this real-life scenario, you'll gain insights into the process of designing and structuring your data tables effectively. Sarah did consider preparing an **entity-relationship** (**ER**) diagram, but her customers didn't want to pay for this as part of the design; they said they would rather see the pages and that she could show them tables of data represented in a view. Sarah did some background reading on ER diagrams before this.

> **Further reading on ER diagrams**
>
> Go to https://learn.microsoft.com/en-us/power-apps/developer/
> data-platform/use-metadata-generate-entity-diagrams to learn more
> about ER diagrams.

Creating tables in Dataverse

To start implementing the data model, Sarah must create three tables: **Incident**, **Incident Type**, and **Injury Type**. These three tables must be implemented to capture and organize the relevant data in our incident management feature. Each table serves a specific purpose: the **Incident** table stores the incident details, the **Incident Type** table categorizes the types of incidents, and the **Injury Type** table captures specific injury classifications. By implementing these tables, Sarah can create a structured data model that facilitates incident management and reporting.

It's recommended to begin with the **Incident Type** and **Injury Type** tables as they serve as reference tables or lists. Since these tables have a simpler structure and no additional fields, ownership can be set to the organization, making the table more efficient without ownership fields.

When creating a table in Dataverse, the system automatically generates a **Name** field, which serves as the primary field and is displayed when interacting with the list. This primary field is sufficient for a basic list view, eliminating the need for additional fields.

Let's dive into the step-by-step process of creating these tables, defining their fields, and establishing the necessary views within Dataverse. This foundation will lay the groundwork for efficient data management and information organization in Sarah's incident management feature.

Type tables

To help you understand how to create a table, we will illustrate creating the **Incident Type** table:

1. Open Power Pages Design Studio and edit the site.
2. Select **Home**, then the **Solution** tab.
3. Open your solution – for example, **Buildapp**, which Sarah created in the previous chapter.
4. Select **+ New | Table | Table**, as shown in *Figure 4.2*:

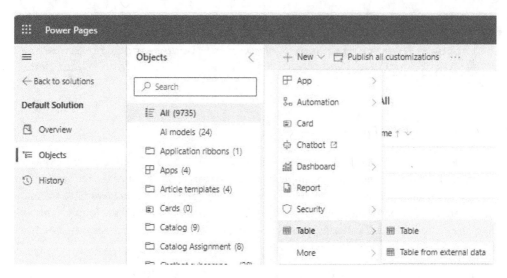

Figure 4.2 – Creating a new table in a solution

Figure 4.2 illustrates how Sarah would create a new table within a solution. We have opened a solution from within Power Pages Design Studio and added a new table. Sarah had to create two new tables for **Incident Type** and **Injury Type** reference data.

> **Tip**
> Working within a solution helps you organize your work better; you can have a solution for each project or feature. You can also add new or existing tables and other objects to work within your solution.

For efficient use of data storage, it is good practice to configure a table that stores reference data records. You can do this by setting **Record ownership** to **Organization** with no additional options enabled, as shown here:

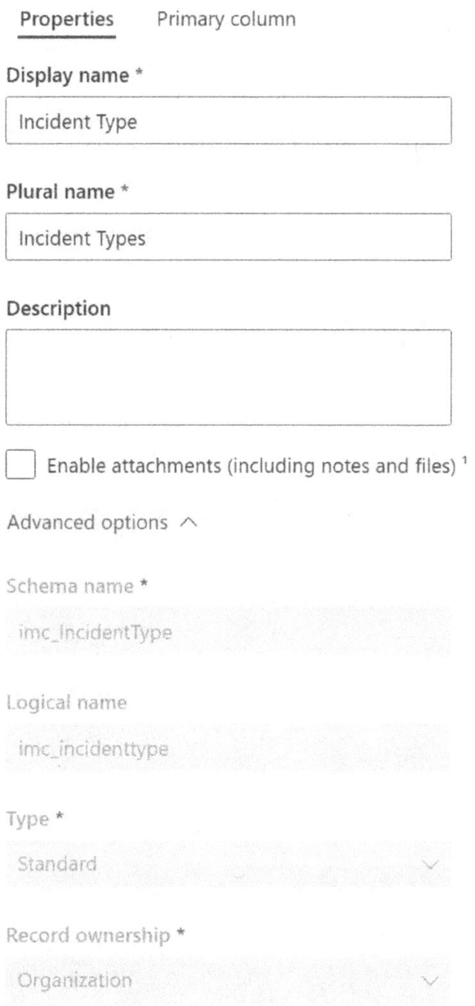

Properties   Primary column

Display name *

Incident Type

Plural name *

Incident Types

Description

Enable attachments (including notes and files) [1]

Advanced options ∧

Schema name *

imc_IncidentType

Logical name

imc_incidenttype

Type *

Standard

Record ownership *

Organization

Figure 4.3 – Configuring a new reference data table for Incident Type

Having created a reference data table called **Incident Type**, repeat this process to create the **Injury Type** table. You can do this by browsing to the **Buildapp** solution, selecting a new table, and entering Injury Type as the table name while providing a similar configuration that you did for **Incident Type**.

The Incident table

Having created the two reference tables, Sarah can now create the **Incident** table. For the **Incident** table, Sarah wants to add the configuration options shown in the following screenshot. Here, Sarah enables attachments by checking the **Enable attachments (including notes and files)** box and selects **User or team** under **Record ownership**. Enabling attachments will allow notes and attachments such as images and files to be added to the incident record. Implementing user ownership will future-proof this table in case Sarah wants to assign user ownership to records at the Dataverse level:

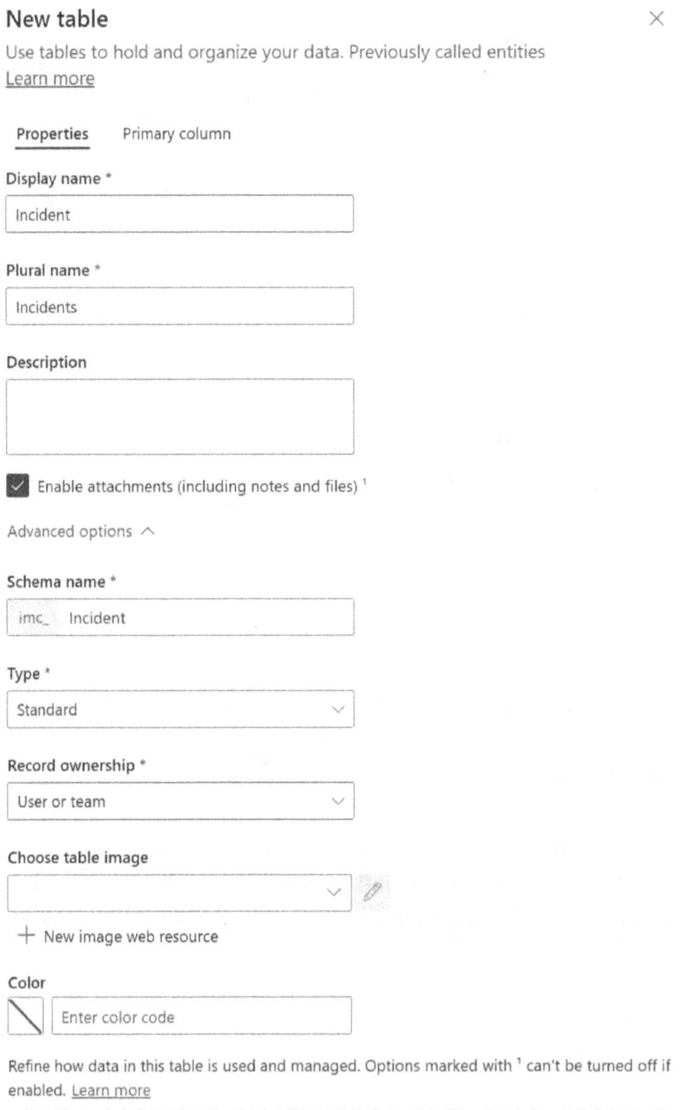

Figure 4.4 – Creating the Incident table

Sarah can also enable activities, as shown in the following screenshot, to enable timeline features such as portal comments, emails, and tasks. This means that emails and tasks can be related to an incident record:

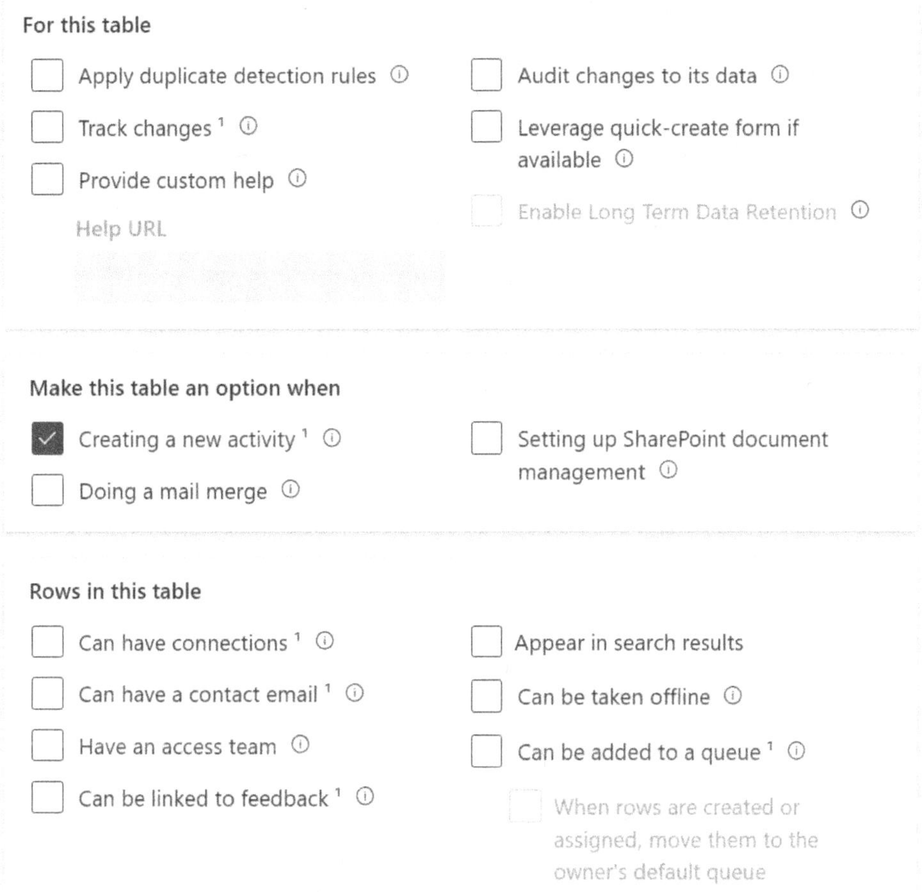

For this table

☐ Apply duplicate detection rules ⓘ   ☐ Audit changes to its data ⓘ

☐ Track changes [1] ⓘ   ☐ Leverage quick-create form if available ⓘ

☐ Provide custom help ⓘ   ☐ Enable Long Term Data Retention ⓘ

Help URL

Make this table an option when

☑ Creating a new activity [1] ⓘ   ☐ Setting up SharePoint document management ⓘ

☐ Doing a mail merge ⓘ

Rows in this table

☐ Can have connections [1] ⓘ   ☐ Appear in search results

☐ Can have a contact email [1] ⓘ   ☐ Can be taken offline ⓘ

☐ Have an access team ⓘ   ☐ Can be added to a queue [1] ⓘ

☐ Can be linked to feedback [1] ⓘ   ☐ When rows are created or assigned, move them to the owner's default queue

Figure 4.5 – Enabling attachments

Now that Sarah has created the necessary tables (**Incident**, **Incident Type**, and **Injury Type**, in the next section, Sarah will work to establish relationships between them to enhance the data model's functionality.

Creating table relationships in Dataverse

Table relationships allow us to link related data across different tables, providing a more comprehensive view of the incident management feature. In the following steps, Sarah will explore how to create these table relationships and leverage the lookup fields to establish seamless connections between the **Incident** table and the reference data tables.

By adding lookup fields in the **Incident** table that go to the **Incident Type** and **Injury Type** reference tables, we can create associations between records. This "many-to-one" relationship enables us to categorize incidents based on their types and link specific injuries to each incident. These relationships enhance data integrity, querying capabilities, and reporting functionalities within the application.

Creating lookups to reference data tables

In this section, Sarah will learn how to create lookup columns on a table, such as **Incident**, which, in turn, will create relationships between the **Incident Type** and **Injury Type** tables in Dataverse. On the **Incident** table, Sarah will create a lookup field to the **Incident Type** table, which creates a many-to-one relationship from the **Incident** table to the **Incident Type** table. A lookup column provides a reference to the **Incident Type** table.

How to create a lookup column

In the **Buildapp** solution, browse to the **Incident** table and add a new column by clicking on **+ New column**. Call this new column Incident Type:

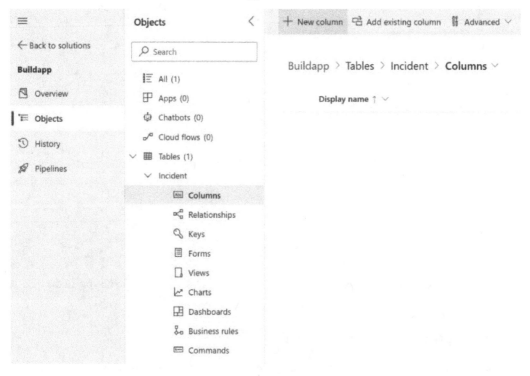

Figure 4.6 – Adding a new column

To complete the new column configuration, follow these steps:

1. In Power Pages Design Studio, select the solution you wish to edit.

2. Browse to the **Tables** tab and select **Incident**.

3. Select **+ New column**, as shown in the preceding screenshot.

4. Set **Display name** to Incident Type.

5. Set **Data type** to **Lookup**.

6. Set **Related table** to **Incident Type**:

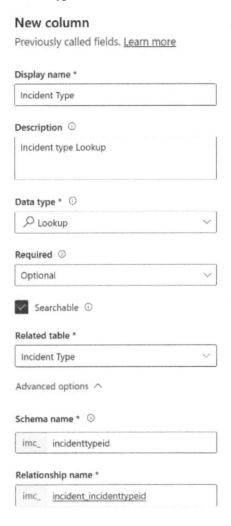

Figure 4.7 – Configuring the Incident Type lookup

The new column configuration window for the **Incident Type** lookup is shown in the preceding screenshot. It's important to follow a consistent naming pattern that makes sense. Sarah will use these names later when selecting relationships while setting table permissions and setting related data in basic forms. In the original Microsoft Dynamics editor, the lookup names would default with a suffix of `id`. It is recommended to follow the widely held practice of renaming your schema name so that it's in lowercase – for example, `incidenttypeid`.

It is also good practice to rename the relationship name so that it's simpler and easier to read – for example, `incident_incidentypeid`. Lastly, adding a description to the column by using the **Description** field can be useful when you want tooltips to appear. Therefore, it is recommended to add meaningful, informative descriptions.

Due to Sarah creating the lookup field, Dataverse will automatically create a many-to-one relationship because it's a lookup column, as shown in *Figure 4.8*:

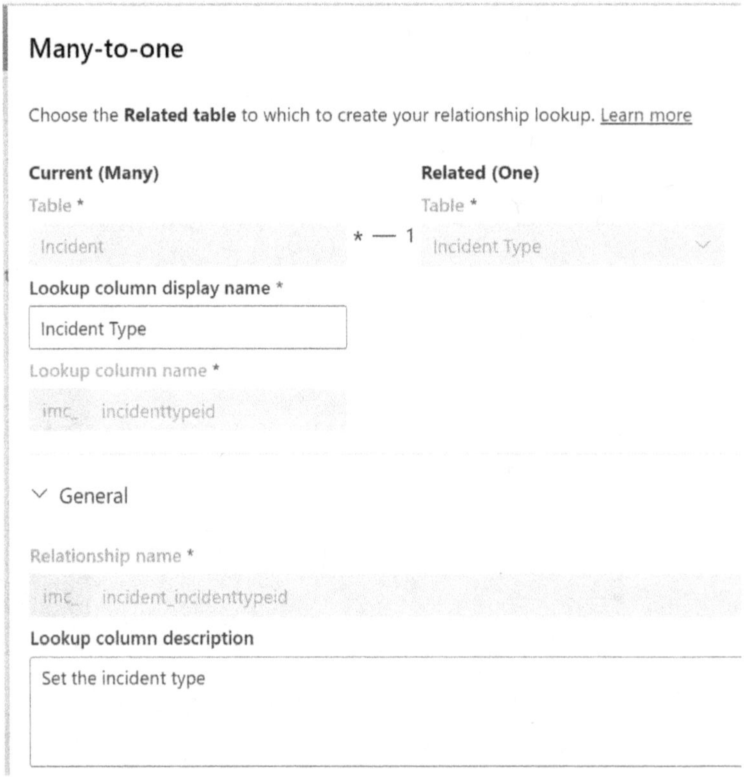

Figure 4.8 – Many-to-one relationship created automatically for Incident Type

Having created a lookup column for **Incident Type**, Sarah wants to repeat this process for the **Injury Type** lookup. Sarah needs to create these lookup-type columns on the **Incident** table so that she has the necessary table relationships for classification, filtering, and reporting.

Creating contact lookups on the Incident table

Sarah also wants to create two lookups on the **Incident** table from the **Contact** table: an **Employee** column and an **Originator** column. These create a many-to-one relationship. Both contact fields will identify people based on an incident.

Employee lookup

In our app, employees are recorded as contact records, so Sarah can add a lookup column based on the incident to the existing **Contact** table and call the **Employee** lookup column. This is useful for employee write-up incidents. Set this to **Business recommended** so that it can be configured as required on a web page:

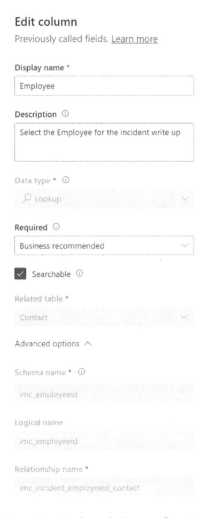

Figure 4.9 – Employee lookup configuration

Originator lookup

Sarah also wants to record the originator – that is, the person who raised and created the incident record on the web page. The originator is the logged-in contact record, so Sarah wants to add a lookup field from the **Incident** table to the existing **Contact** table and call the **Originator** lookup column. This is used by table permissions to give the creator delete permission and also to record who authored the incident. In the next chapter, Sarah will configure this field so that it's automatically filled in when the insert record is saved:

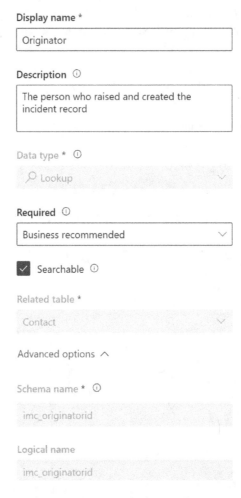

Figure 4.10 – Originator lookup configuration

In the next section, Sarah will show you how to implement many-to-many relationships and why they are used.

Many-to-many (N:N) relationships for people contacts

Sarah needs subgrid lists of contacts in the incidents form. These subgrids are designed with many-to-many relationships between the **Contact** table and the **Incident** table. Sarah uses many-to-many relationships so that a contact can appear on many different incident records. This enables users to add and remove contacts to/from any incident. Sarah will create three many-to-many relationships for contact subgrids on an incident for **witnesses**, **people notified**, and **people involved**. To create a many-to-many relationship, browse to the **Incident** table's **Relationships** tab and choose a new **Many-to-many** relationship, as shown here:

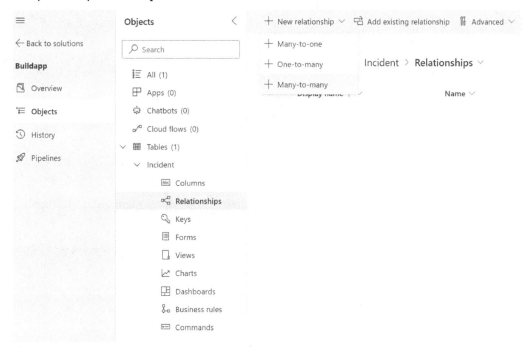

Figure 4.11 – Creating a new many-to-many relationship on the Incident table

Configure the many-to-many relationship for witnesses on the **Incident** table:

1. In Power Pages Design Studio, select the solution you wish to edit.
2. Browse to the **Tables** tab and select **Incident**.
3. Select the **Relationships** tab, as shown in the preceding screenshot.
4. Select + **New relationship**, then + **Many-to-many**.
5. Select the **Contact** table for the **Related** table.
6. Set **Relationship name** to incident_witness_contact.
7. Set **Relationship table name** to incident_witness_contact.

When Sarah selects a new many-to-many relationship, the following configuration form opens, showing the current table – that is, **Incident**. To create a relationship with the existing **Contact** table, set the related table to **Contact** and enter a relationship name. Sarah has set this to incident_ witness_contact:

Many-to-many

Choose the **Related table** to create your relationship. Learn more

Current (Many)		Related (Many)
Table *		Table *
Incident	* — *	Contact ⌄
Relationship name *		
imc_ incident_witness_contact		
Relationship table name *		
imc_ incident_witness_contact		

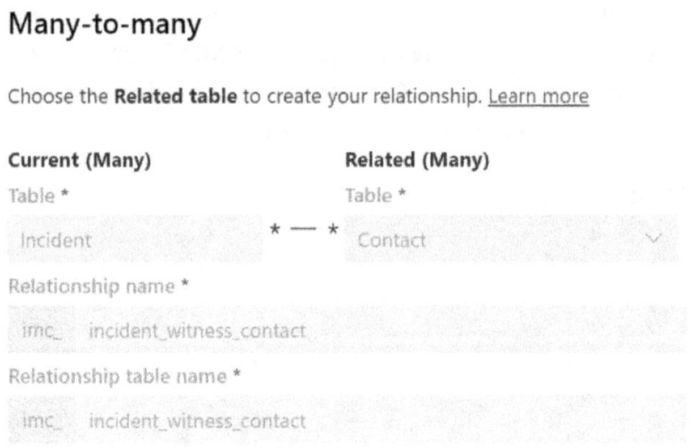

Figure 4.12 – Witnesses many-to-many relationships

Repeat this for the other two relationships to create three many-to-many relationships to enable subgrids on the incident forms.

Create the following three many-to-many relationships. These will be used later in this chapter when we configure the different forms:

1. **Witnesses**
2. **People Involved**
3. **People Notified**

This will result in the following many-to-many relationships being available on the **Incident** table:

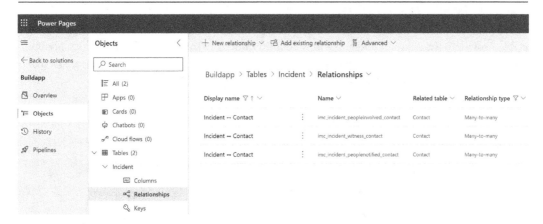

Figure 4.13 – Many-to-many table relationships on the Incident table

Having created the related lookup columns for the many-to-one table relationships and completed the many-to-many table relationships, Sarah will create the remaining columns to be used by our incident management feature.

Creating table columns

Now, Sarah will create the other columns that will used in the feature. She will specify types from the planned data model as she won't be able to change the type once the column is created. These columns will be placed on the forms on the web pages.

> **Tip**
> To further understand column types, read the following documentation from Microsoft Learning: https://learn.microsoft.com/en-us/power-apps/maker/data-platform/create-edit-field-portal#column-data-types.

Sarah should have a meaningful data collection process in place for the incident on the form. Let's look at some typical columns that should be used on the incident form. She will use these to capture and report on the different incidents.

Columns can be created by following the process described in the *How to create a lookup column* section. Browse to the solution and add a new column under the **Incident** table. The schema name is automatically set when the field name is entered. These schema names can be overwritten, just like Sarah did with the `imc_managingaccountid` column, as shown here:

Column Display Name	Schema Name	Form Type
Managing Account	`imc_managingaccountid`	Lookup
Project	`imc_projectid`	Lookup
Date/Time	`imc_datetime`	DateTime
Notified Date/Time	`imc_notifieddatetime`	DateTime
Accepted Treatment	`imc_acceptedtreatment`	Boolean
Any Injuries	`imc_anyinjuries`	Boolean

Table 4.2 – Table of incident columns to add

Setting the date of the incident column

The **Date** type defaults to the user locale. This is usually preferred as that way, it's always displayed in the user-preferred time format and according to the user's locale. Sarah has selected the date and time format as she needs to know the exact time of an incident, especially if this could help form notes for legal action.

> **Tip**
>
> Further reading on the DateTime data type can be found at `https://learn.microsoft.com/en-us/power-apps/maker/data-platform/create-edit-field-portal#date-time`.

Global choices

We need lists as dropdowns on web pages. In the preceding table, we have two columns that must be created as choice columns: **Occurrence** and **Type**. In Dataverse, these are called choice types; Sarah has the option of creating a global or local table choice. When creating choice fields, it is often best to select global choice so that the choice can be used in other tables as it could well be used in other tables.

Create a new column for **Occurrence** and select a data type of your choice. Once you've done this, the choice configuration shown in the following screenshot will appear. Sarah hasn't added a **default choice** for **Occurrence**, which means that the field will be blank on page load and the user will need to select a value from the drop-down choices field. In the choice configuration form, Sarah can create a global choice by selecting **Yes** for **Sync with global choice?**, as shown in *Figure 4.15*:

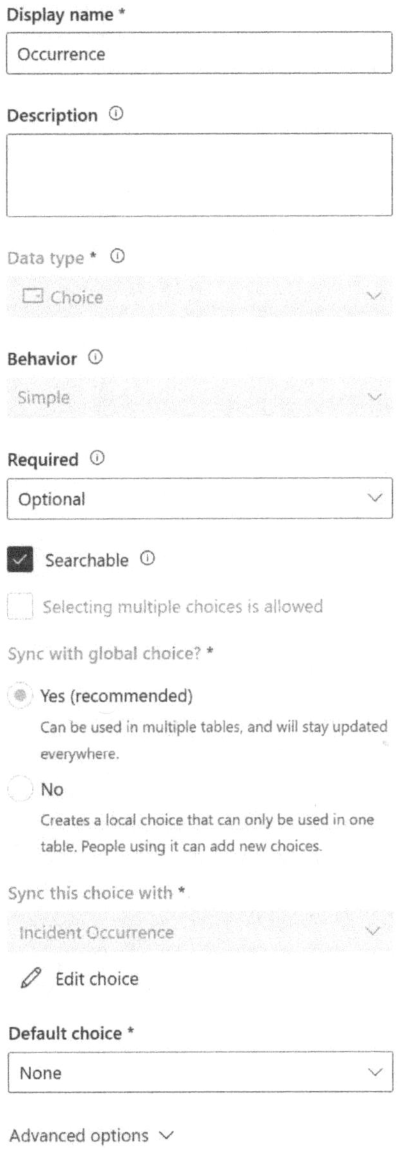

Display name *

Occurrence

Description ⓘ

Data type * ⓘ

☐ Choice   ⌄

Behavior ⓘ

Simple   ⌄

Required ⓘ

Optional   ⌄

☑ Searchable ⓘ

☐ Selecting multiple choices is allowed

Sync with global choice? *

⦿ Yes (recommended)
    Can be used in multiple tables, and will stay updated
    everywhere.

○ No
    Creates a local choice that can only be used in one
    table. People using it can add new choices.

Sync this choice with *

Incident Occurrence   ⌄

✎ Edit choice

Default choice *

None   ⌄

Advanced options ⌄

Figure 4.14 – Global choice configuration for the Occurrence column

To modify or add new choice values, Sarah must click **Edit choice**, which opens the following configuration form. In these choice fields, she sets **Display name** to **Incident Occurrence**:

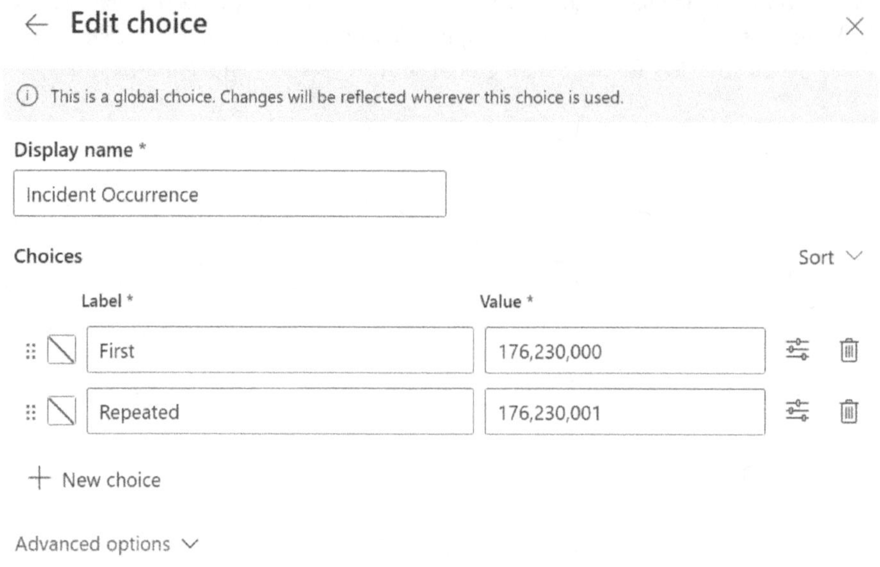

← **Edit choice**   ✕

ⓘ This is a global choice. Changes will be reflected wherever this choice is used.

Display name *

Incident Occurrence

Choices   Sort ∨

Label *	Value *		
First	176,230,000		
Repeated	176,230,001		

＋ New choice

Advanced options ∨

Figure 4.15 – Occurrence choice values

Note that Sarah leaves the **Choices** values as the defaults that Dataverse generates, as illustrated in the preceding screenshot. This is fine to do.

The Form type choice

The **Form type** choice here is for an **Incident** type of **Employee write up**, which is a common incident type within the construction industry. Sarah wants this choice of form type so that she can redirect the user to the correct incident form. Sarah plans to create two different forms, one for **Incident** and one for **Employee write up**, which we will see her implement later in this book. These two form types will be used to filter records, organize our data, and filter reports.

Follow these steps to create the **Form type** choice field:

1. In Power Pages Design Studio, select the solution you wish to edit.
2. Browse to the **Tables** tab and select **Incident**.
3. Select the **+ New** column.
4. Set **Display name** to Form Type.
5. Set **Data type** to **Choice**.
6. Set **Required** to **Business required**.

7. Create a global choice named `Incident Construction choice`.

8. Select **Yes** for **Sync with Global choice?** regarding **Incident Construction choice**.

9. For global choice, select **Edit choice** to manage the actual choices. Add the **Incident** and **Employee Write-Up** options, as shown in *Figure 4.18*.

10. Select **Incident** as the default value of the **Form Type** choice:

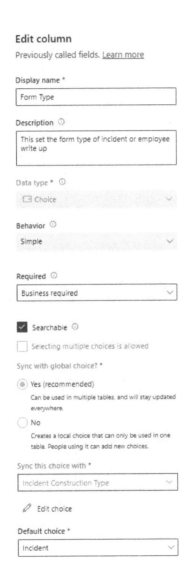

Figure 4.16 – The Form Type column's configuration with Incident as the default value

Steps 8 and *9* create a global choice as Sarah creates the **Form Type** choice column in the **Incident** table. Once she's created the global choice, Sarah can add the choice options shown in the following screenshot. Once these choice values have been added and saved, Sarah can select **Incident** as the default value:

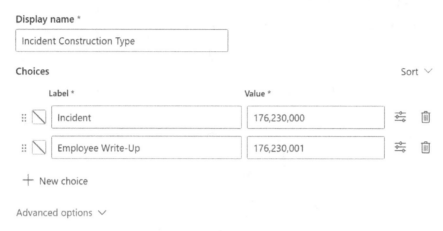

Figure 4.17 – Form Type column choice values

Multi-line text fields

Many of the text fields to be created on the **Incident** table have been designed to be in multi-line text format, which means that they aren't single lines but blocks of text in paragraphs on a web page. The **Description** column is an example of this. When selecting multi-line text fields, Sarah must specify the length in characters. In this case, Sarah has selected 2,000 characters, which is the default setting. Note that the larger the value, the greater the amount of database storage that's allocated to the table. This affects the cost of the systems as Microsoft charges storage by allocated space, regardless of whether the rows contain data or not. Please note that you can adjust the size of these fields by either reducing or increasing them.

The Status Reason field

When creating a new table, the system creates some system fields, including a **Status Reason** field. This special field can be used to organize and group records in a view. It's common to use the **Status Reason** field to trigger automations. Sarah will need to view separate views for open incidents, completed incidents, and canceled incidents.

> **Note**
> Note that on Power Pages, the **Status Reason** field cannot be displayed on a form, and therefore users can't set it on a web page. However, you can set the values in a process or Power Automate flow.

Later in this book, we will see Sarah implement a cancellation process that will set **Status Reason** to **Cancelled**. So, Sarah needs a **Cancelled** option. To do this, Sarah can browse to the **Status Reason** field, edit it, and add a new option called **Cancelled**.

A system-generated value is provided under **Inactive** for **Status Reason**. Sarah will relabel this inactive option Completed, as shown here:

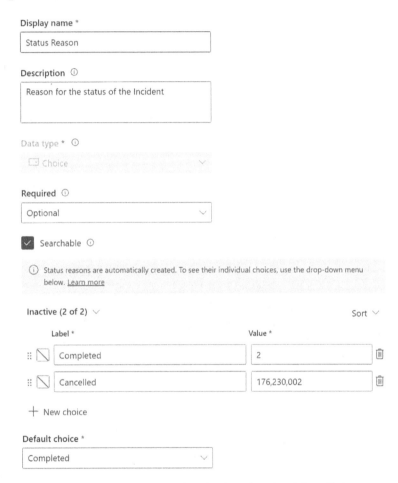

Figure 4.18 – Inactive values for Completed and Cancelled

With that, Sarah has set both **Completed** and **Cancelled** to **Inactive**. She's done this because the Dataverse system will lock records and prevent them from being writeable once **Status Reason** is set to an inactive state. Sarah must do this to ensure changes aren't made to a completed record.

Sarah will also relabel **Active** to **Open** as it's an open incident. She's done that so that she has a view of open incidents:

Figure 4.19 – Active values

So far, we've learned how to plan the development of a feature, as well as how to create tables and create relationships between these tables. Then, we learned how to implement fields, as well as how to decide on what data types and formats to use when creating fields. Having created and implemented the configuration for the various incident columns, in the next section, we will see Sarah configure existing views and create new views as needed in her app.

Views

Views are defined lists of ordered columns that are configured via sorting and filtering. Earlier, Sarah modified **Status Reason** to enable filtered views of incidents. Now, we will see how Sarah configures these views. When creating tables, the system generates views. It is important to modify these existing views and create new ones.

> **Further reading**
>
> You can read more about views in the Microsoft Learning documentation: `https://learn.microsoft.com/en-us/power-pages/configure/data-workspace-views#view-designer`.

Active views are the default views and are created automatically when creating a new table. The **Active Incidents** view is the default view for incidents and is created alongside the **Name** and **Created on** columns. The **Active Incidents** view is created with a filter of `status eq active`. This filter will show any open incidents – that is, a list of records that have **Status Reason** set to **Open**. The idea here is to add more meaningful columns so that we can use this view in Power Pages. Let's see how Sarah does this.

How to edit an existing view

To modify the view, Sarah must work on the **Buildapp** solution, which she will access from Power Pages Design Studio:

1. In Power Pages Design Studio, select the **Buildapp** solution.
2. Browse to the **Tables** tab and select **Incident**.
3. Select the **Views** tab of the **Incident** table.
4. Select the **Active Incidents** view, as shown in *Figure 4.21*.
5. Click on the ellipsis to edit the view in a new tab.
6. Opens the view designer, as shown in *Figure 4.22*. This is where Sarah will edit the view:

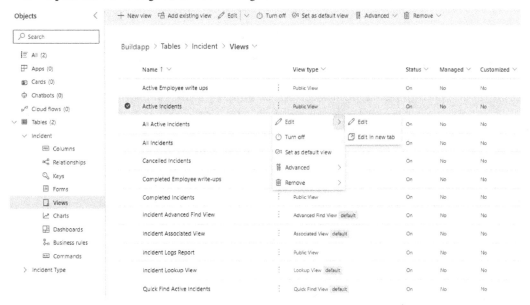

Figure 4.20 – Editing the Active Incidents view

For the most part, Sarah will always want to configure the default view so that it has more columns and sometimes change the configured sorting. Having gone to the **Active Incidents** view, she modifies this default view in the designer, where she can manage columns, add and remove columns, and modify sorting and column widths.

Modifying the Active Incidents view

To make this default view more useful and present an informative view relevant to incidents, Sarah will add four columns to the custom fields we created earlier in this chapter.

Sarah adds the following columns, which are highlighted in the rectangular box in *Figure 4.22*, by dragging and dropping the columns into the view:

1. **Date of Incident**
2. **Form Type**
3. **Incident Type**
4. **Description**:

Figure 4.21 – The Active Incidents view's columns in the view designer

The system-generated view of active incidents will be sorted by name, but that's not very useful in a default view. It's common to have the default open incidents view sorted by **Date of Incident**, with the newest at the top. Sarah wants this view to only show incidents and not show employee write-ups, so she adds filter-by-form-type criteria, as shown in *Figure 4.23*. After making these changes, she saves and publishes her work:

Figure 4.22 – Active Incidents sorting and filtering configuration

Having modified the **Active Incidents** view so that it shows open records of form-type incidents, Sarah now wants a similar view for active employee write-up incidents.

The Active Employee write ups view

The best way to create a similar view is to open an existing view and save it as a new view. To do this, Sarah edits the **Active Incidents** view and saves it as **Active Employee write ups**. Then, she modifies **Filter by …** so that it includes **Type is 'Employee Write-Up'**:

Figure 4.23 – The Active Employee write ups filter

Having completed the two open records for each form type of incident, in the next section, Sarah will create views for the **Completed** and **Cancelled** views.

The Completed and Cancelled views

Earlier in this chapter, we saw how Sarah modified **Status Reason** for the two inactive states of **Completed** and **Cancelled**. Here, we will see how Sarah creates and configures these views. As mentioned previously, it's quicker to open an existing view and save it, though Sarah can also create a new view and configure its columns and filters.

The Completed views

As before, we will edit the **Active Incidents** view, save it as **Completed Incidents**, and apply the following configurations:

Figure 4.24 – Configuring the Completed Incidents view

Note that we set **Filter by …** to **Status Reason is 'Completed'** and **Form Type is 'Incident'**. Sarah saves and publishes her work. Then, she saves this as **Completed Employee write ups** and configures the rest of the fields, just like she did for the **Completed Incidents** view:

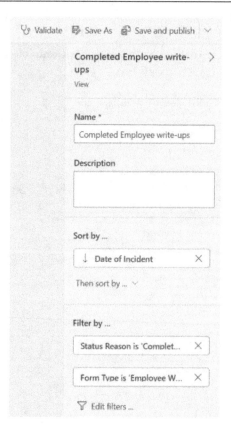

Figure 4.25 – The Completed Employee write-ups view

The Cancelled views

To create the **Cancelled Incidents** view, Sarah opens the existing **Completed Incidents** view and saves it as **Cancelled Incidents**. After, she modifies the **Status Reason** filter and sets it to **Cancelled**. Then, she saves and publishes her work.

To create the **Cancelled Employee write ups** view, Sarah opens the existing **Completed Employee write ups** view, saves it as **Cancelled Employee write ups**, and then modifies the **Status Reason** filter and sets it to **Cancelled**. Then, she saves and publishes her work.

Having completed the incident views, Sarah needs to create views in the **Contact** table. These contact views will be used in the **Incident** table lookups on the web page forms to help users filter and find their contacts, thereby enhancing the user experience.

Contact table views

Sarah wants to create some custom views to organize her contacts into filtered lists, which help users select contacts from employees or customers or **My Crew** or vendors lists.

Creating Boolean columns to filter views

To have a new view filtered by a contact type, Sarah will need to create new columns to enable these views – columns consisting of a **Boolean** type of **Yes/no** field for each new view. Sarah uses a Boolean field to identify the type of contact as this allows us to have a contact with multiple types – for example, a contact could be **My Crew** as well as **Employee**. So, an employee who is a member of **My Crew** will have **Is Employee** set to **Yes** and **Is My Crew** set to **Yes**.

Sarah creates the following **Yes/no** columns:

1. **Is Contractor**
2. **Is Customer**
3. **Is Employee**
4. **Is My Crew**
5. **Is Vendor:**

Construction > Tables > Contact > **Columns** ∨

Display name ↑ ∨		Name ∨	Data type ∨
Is Contractor	⋮	imc_iscontractor	⬭ Yes/no
Is Customer	⋮	imc_IsCustomer	⬭ Yes/no
Is Employee	⋮	imc_IsEmployee	⬭ Yes/no
Is My Crew	⋮	imc_ismycrew	⬭ Yes/no
Is Vendor	⋮	imc_isvendor	⬭ Yes/no

Figure 4.26 – Yes/no columns to add to the Contact table

Next, Sarah must create corresponding views for each of these options to allow filtering to occur when contacts are being selected. Later in this book, Sarah will use these views on forms and pages.

She creates a new view, one for each new Boolean column. Then, she configures the filter for each view with the corresponding column equal to yes. So, the **Customers** view has a filter of **Is customer equal yes.**

A quick way of doing this is to open the **Active Incidents** view, save it as **Customers**, and then modify the filter. She repeats this process for each of the new views:

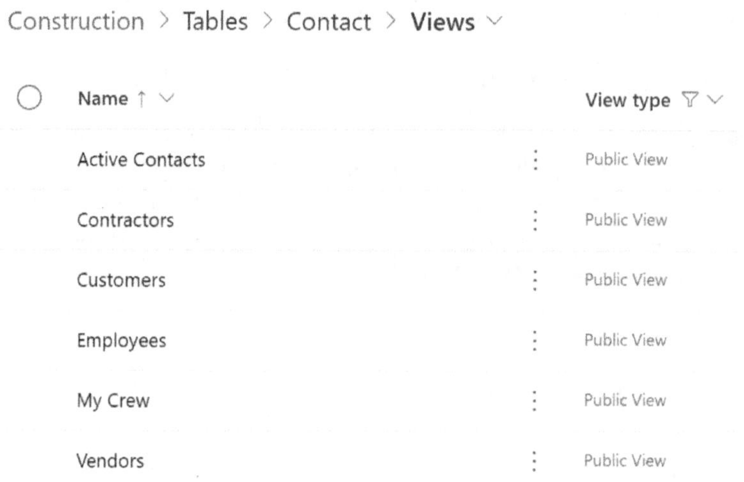

Construction > Tables > Contact > **Views** ⌄

	Name ↑ ⌄		View type ▽ ⌄
○	Active Contacts	⋮	Public View
	Contractors	⋮	Public View
	Customers	⋮	Public View
	Employees	⋮	Public View
	My Crew	⋮	Public View
	Vendors	⋮	Public View

Figure 4.27 – These views have been configured with the Yes/no column

Now that Sarah has her contact views, she's ready to dive into the **Report** view.

The Report view

Earlier, in our planning stage, we discussed list pages, which list incident records. Users use the list page to organize, filter, search, and access incident records. A useful pattern is a view with most or all columns available. This is called a **Report** view.

Sarah will create a **Report** view with nearly all the columns on the forms set in a meaningful order. Then, on a list page, she'll implement the **Download** function, something she'll implement in the next chapter. This pattern enables users to download incident data from the list page.

Now that Sarah has created a data model consisting of tables, columns, and views, she needs to design the forms that will be used on the Power Pages website.

Designing Dataverse forms

In this section, Sarah will create the forms that will be used on the website. She will design an insert form and two edit forms for each form type – the **Incident** form type and the **Employee write-up** form type. She will add these fields to these forms and add subgrid controls for the related many-to-many contacts she created earlier before configuring the subgrids.

> **Tip**
> To learn more about how to add forms to pages, go to `https://learn.microsoft.com/en-us/power-pages/getting-started/tutorial-add-form-to-page`.

Insert forms

First, Sarah will create an insert form; she will work on this in the Power Pages **Buildapp** solution:

1. In Power Pages Design Studio, select the **Buildapp** solution.
2. Browse to the **Tables** tab and select **Incident**.
3. Select the **Forms** tab on the **Incident** table.
4. Add a new main form called `Portal`, as shown in *Figure 4.29*.
5. This will open the form designer, as shown in *Figure 4.30*:

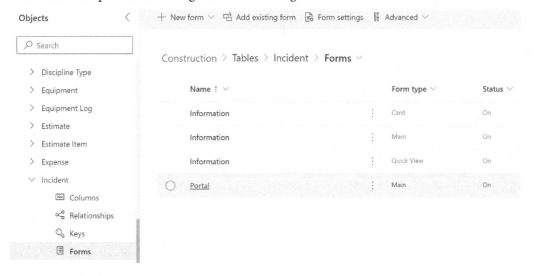

Figure 4.28 – Adding a new form

In the form designer, Sarah adds a new tab, gives it a label of **New Incident**, names it **incidentinsert**, and adds **Name**, **Date of Incident**, and **Form Type** as required fields, as shown here:

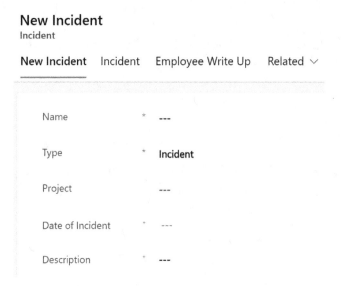

Figure 4.29 – Creating a tab

Sarah can configure the tab format as she wishes. For the insert form, a **1 column** tab with a **1 column** section is good for a simple form. Now, let's create the edit forms.

Edit forms

Edit forms can also be used for edit and read-only forms. In this section, we'll cover the **Incident** form type, the **Employee write-up** form type, and the subgrid's configuration. We'll also learn how Sarah configures views on a subgrid.

Sarah needs two edit forms – one for **Incident** and another for **Employee write-up**. They will both have different fields on the web, so Sarah will use a form tab, one for each incident type form.

The Incident form tab

To create the **Incident** form tab, Sarah adds a new tab called **Incident**. This form tab will be used to present a web page form for incidents and will be used for edit and read-only forms of the **Incident** form type.

Form tabs and their sections can be configured in terms of their format, and four columns can be specified. Sarah's preference for such a form with lots of columns is to have a **2 column** section. In the **Incident** tab, create such a column section and drag the fields onto the form, as shown here:

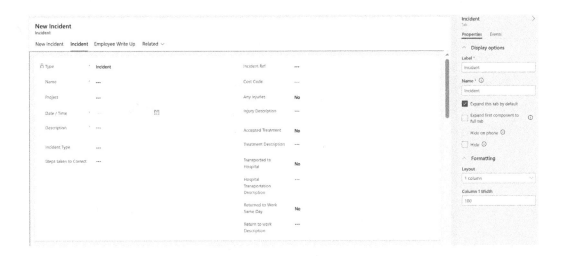

Figure 4.30 – Editing the New Incident form tab

Note that on mobiles, this will display as **1 column** automatically.

The Employee write-up form tab

To create the **Employee write-up** form tab, Sarah will add a new tab called **Employee Write Up**. This will be used for edit and read-only forms of the **Employee write up** form type. For design consistency, Sarah will use the same design she did previously and apply a **2 column** section. As shown in *Figure 4.32*, add a **2 column** section to the **Employee Write Up** tab and drag the **Employee Write Up** fields onto the form:

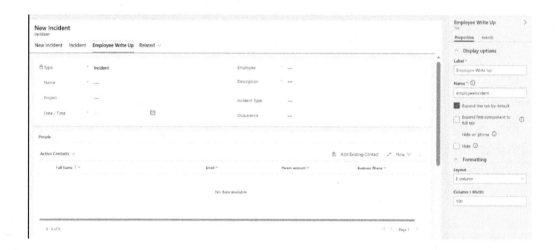

Figure 4.31 – Adding a 2 column section

Having added the fields for the **Employee Write Up** form in the **2 column** section for the fields, Sarah needs to add a section under that that will contain the two people subgrids. We want to give the subgrids as much space as possible and use all the horizontal space of the form. We can do this with a **1 column** section.

Subgrid configuration

A subgrid contains a list of related records. Here, Sarah wants to have subgrids so that people involved in the incident can be listed.

These subgrids contain the people involved in the **Employee Write Up** process. Sarah adds a **1 column** section and calls the section People, as shown in *Figure 4.32*. She then configures the **People** section, as shown in *Figure 4.33*:

Figure 4.32 – Configuring the People section

Now that this section is ready, Sarah wants to create the necessary subgrids. To add a subgrid via the form designer, Sarah drags a subgrid from **Components** onto the **People** section. The first subgrid she calls is **Witnesses**; this will use the **Witnesses** contact's many-to-many relationships that were created earlier in this chapter. Configure the **Witnesses** subgrid, as shown in *Figure 4.34*, so that it displays witnesses. Sarah does this like so:

1. Enable the **Show related records** option so that it will filter and only show records related to this incident.

2. Select **Witnesses** under **Table** to show witness contact relationships.

3. Select the default view for this subgrid:

Figure 4.33 – Configuring the Witnesses subgrid

Note that Sarah is using the **Active Contacts** view, having completed the **Witnesses** subgrid, to create the **People Involved** subgrid. To do this, Sarah drags a subgrid component under the **Witnesses** subgrid and calls it **People Involved** while using the configuration shown in *Figure 4.35*.

Sarah configures the subgrid with the **People Involved** contact relationships that she created earlier in this chapter:

Figure 4.34 – Configuring the People Involved subgrid

In the form designer, the two subgrids will look as follows:

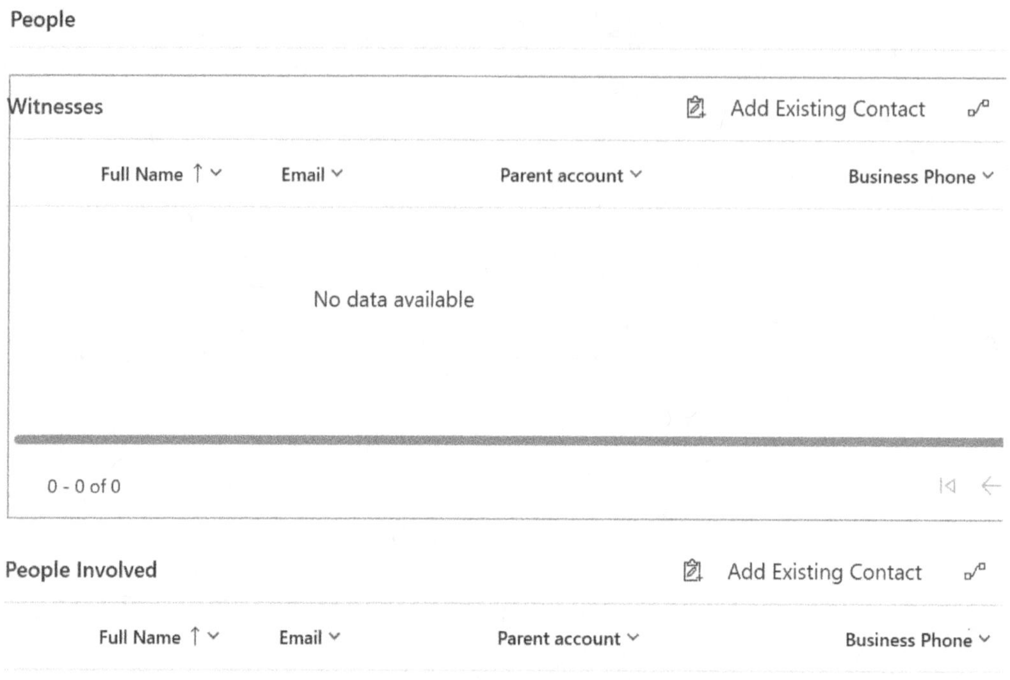

People

Witnesses   📋 Add Existing Contact   ⬈

Full Name ↑ ∨	Email ∨	Parent account ∨	Business Phone ∨

No data available

0 - 0 of 0 |◁ ←

People Involved   📋 Add Existing Contact   ⬈

Full Name ↑ ∨	Email ∨	Parent account ∨	Business Phone ∨

Figure 4.35 – Subgrids containing related contacts

Having implemented the various people subgrids, Sarah needs to configure the selectable views of each subgrid to make finding people easier and finish configuring the subgrid.

Configuring views on a subgrid

Earlier in this chapter, we saw Sarah create contact views to help users search for and filter contacts when selecting contacts to add. This is useful if there are many contacts. Sarah will configure these views on both subgrids, as shown here:

Figure 4.36 – Selected views on the contact-related subgrids

Further reading on using subgrids in Power Pages

Go to `https://learn.microsoft.com/en-us/power-pages/configure/configure-basic-form-subgrid` for more information on how to use subgrids in Power Pages.

Now that Sarah has completed the subgrid and form fields, she wants to add a timeline control at the bottom of the form.

Timeline control

A timeline control enables attachments and related activities such as notes, tasks, and emails. It lists related activities and allows activity records to be created and edited.

> **Tip**
>
> You can learn more about timeline controls at `https://learn.microsoft.com/en-us/power-apps/maker/model-driven-apps/set-up-timeline-control`.

While still in the **Employee Write Up** form tab, from **Components**, Sarah drags a timeline control under the subgrids and configures it. Sarah also puts it into its own **1 column** section and labels it **Timeline and notes**:

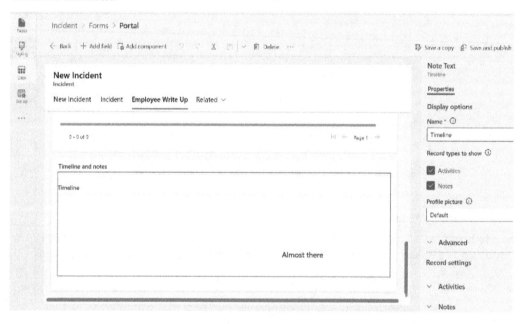

Figure 4.37 – Timeline control

In this section, we learned how to design Dataverse forms. This included inserting and editing forms. Finally, we covered timeline control. Now, let's summarize what we learned in this chapter.

Summary

This chapter marks a significant step in Sarah and Brenda's journey to enhance Rob the Builder's digital infrastructure. By focusing on the development of an incident management system, they addressed a critical need within the construction business – that is, ensuring compliance with industry regulations and enhancing operational efficiency. Through careful planning, thoughtful data modeling, and strategic role assignments, they lay the groundwork for a robust, user-friendly system. Their collaborative approach ensured the solution was not only technically sound but also perfectly aligned with the business's unique requirements. This chapter sets a strong foundation for the actual implementation of the feature, preparing Sarah to bring her technical expertise and Brenda's operational insights to life in the subsequent development stages.

In this chapter, we explored the essential aspects of planning a feature, including how to design a data model and establish table relationships. We also delved into the significance of web roles for ensuring security within our application. Understanding data types in Dataverse, such as lookup types, choice types, and status types, enabled us to structure our data effectively.

Additionally, we learned about the process of creating and configuring views, which enhances data organization and retrieval. Designing and implementing forms, along with subgrids, provided us with a user-friendly interface to interact with the data. These components are vital for a comprehensive and functional feature implementation.

Throughout this chapter, we gained valuable insights into how to plan and execute various elements that contribute to a well-designed and efficient feature, empowering us to create impactful solutions using the Microsoft Power Platform.

We learned the importance of planning a feature, including the design of a data model, table relationships, and the use of web roles for security. This foundational step ensures we have a well-structured and secure application. We gained knowledge about different data types in Dataverse, such as lookup types, choice types, and status types. Understanding these data types allows for accurate data representation and effective data management.

We also covered how to create and configure views, which enable efficient data organization and retrieval. We learned how to leverage views to enhance the user experience and streamline data access.

Then, we explored the design and implementation of forms, including the use of subgrids. We also discovered how to create user-friendly interfaces that facilitate data input and interaction, improving overall usability.

By grasping these concepts and techniques, you are equipped with the knowledge and skills to effectively plan, design, and implement features using Microsoft Power Platform, enhancing your ability to create robust and user-centric applications.

In the next chapter, we will learn how to configure table permissions and Power Pages security. This will give users secure role-based access to the necessary incidents and their related tables.

Table Permissions and Security

In the previous chapter, Sarah, tasked with enhancing Rob the Builder's **incident management system (IMS)**, explored the implementation of two key web roles: **Back Office** and **Foreman**. This was a pivotal step in her journey to bolster the project's security framework. Power Pages utilizes a comprehensive security model to protect business information on public-facing websites. This model incorporates essential components such as site visibility, authenticated users, web roles, table permissions, and page permissions to regulate access and maintain data integrity.

Sarah's approach to enabling access to Dataverse records involves configuring table permissions. She's learning to associate these permissions with the web roles, which are crucial for data security and providing **role-based** security access to specific tables within the Incident Management feature, including **Notes**, **Incidents**, and **Incident Types**.

Additionally, Sarah will examine different access types in table permissions to gain a comprehensive understanding of their application based on web roles.

> **Tip**
> For further reading, follow the link to *Microsoft Learn*: `https://learn.microsoft.com/en-us/power-pages/security/table-permissions`

This chapter will cover the following topics:

- Exploring how table permissions enable access to Dataverse records in Power Pages, with role-based security access
- Configuring table permissions
- Child access permissions
- Introduction to the architecture of table permissions

By understanding and implementing these table permissions, Sarah will gain better control over access and data integrity within the Power Pages application.

User story – table permissions for access to Dataverse

To enable access to Dataverse records in Power Pages, Sarah configures table permissions and links them to web roles, used by forms, lists, Liquid, and other components. In the following sections, Sarah learns how to create table permissions to give users access to the tables created in *Chapter 4*.

Sarah needs to configure table permissions for **Read**, **Create**, **Write**, **Delete**, **Append**, and **Append to** access to records in the tables.

Configure table permissions in the Power Pages studio as follows:

1. Open the Power Pages studio:

    I.   Browse to `https://make.powerpages.microsoft.com/`.

    II. Select your environment.

    III. Select your active site and click **Edit**, which opens the Power Pages studio.

2. Select the **Set up** tab, then select **Table permissions**, as shown in *Figure 5.1*.

3. Developers can add and edit **Read**, **Create**, **Write**, **Delete**, **Append**, and **Append to** table permissions:

Figure 5.1 – Table permissions

> **Note**
>
> It is best to always enable **Table permissions** for all components including for anonymous access, where you should use the anonymous web role.

In the upcoming subsections, Sarah will be learning about all the permissions that she can use and delve into the nuances of configuring different types of permissions in Power Pages, understanding their specific applications and impacts on data security and accessibility. This journey through various permission settings is critical for ensuring that only the appropriate users have the right level of access to the data within the Incident Management feature.

Self-access permissions

These provide access directly to the logged-in user through its direct contact relationship. An example can be seen in the table permission on **Contact** enabling the logged-in contact to edit their own contact record, as shown in *Figure 5.2*:

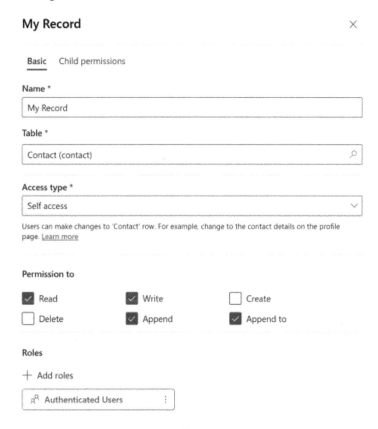

Figure 5.2 – Self access permission

The **Name** value is descriptive, showing that it is access to the record of the logged-in user. The **Table** value specifies that it has permission on the **Contact** table, as the logged-in user is a **contact** record. The **Access type** value is **Self access**, meaning access by logged-in user contact. Each permission is enabled with a checkmark and then a list of web roles that have this permission – in this case, the authenticated system-provided web role.

Global access permissions

Global access means access is not filtered by any table relationship to the logged-in user. This is useful for reference data or static data and for the **Reference data** table of the **Incident Type**; typically, reference data tables only need **Read** and **Append to** access. Configure both the web roles of **Foreman** and **Back Office**, as shown in *Figure 5.3*. As explained in *Chapter 4*, **Append to** will allow users to append **Incident type** lookup records to an incident record:

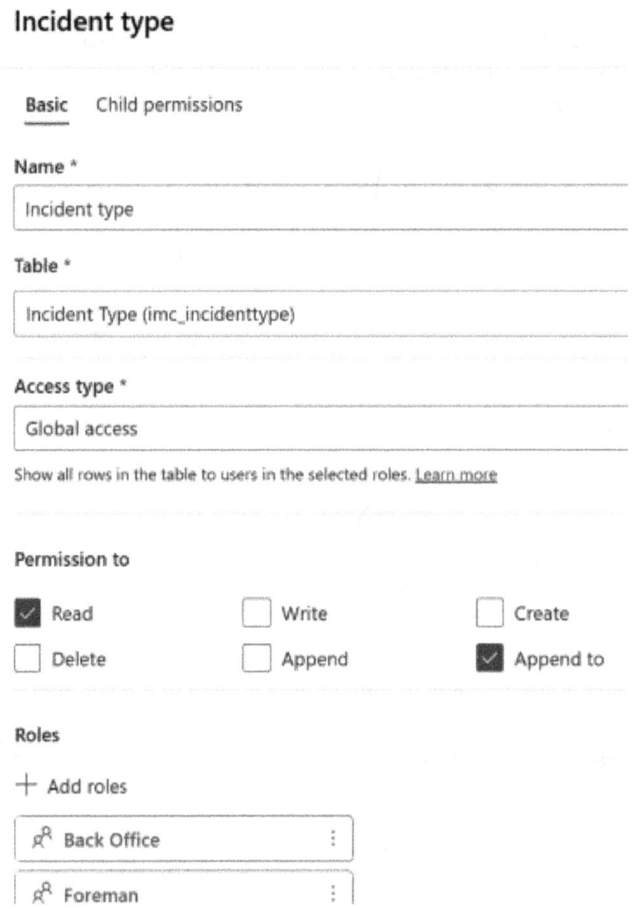

Incident type

Basic   Child permissions

Name *

> Incident type

Table *

> Incident Type (imc_incidenttype)

Access type *

> Global access

Show all rows in the table to users in the selected roles. Learn more

Permission to

☑ Read   ☐ Write   ☐ Create

☐ Delete   ☐ Append   ☑ Append to

Roles

+ Add roles

> 👤 Back Office   ⋮
>
> 👤 Foreman   ⋮

Figure 5.3 – Global access table permission

Contact access permissions

Sarah was briefed that an incident record should have specific permissions to its creator, and she needs to allow **Write** and **Delete** permissions to an incident record only by the person who created it. So, this will be done at the **Access type** field of **contact** and will filter data according to the contact relationship to the table of the logged-in user, created earlier in *Chapter 4*.

To create and configure a table permission for **Incident Originator**, proceed as follows:

1. Open the Power Page studio.
2. Select the **Set up** tab, and then select **Table permissions**.
3. Select the **+ New permission** button to open the form as shown in *Figure 5.4.*
4. Configure the **Name** field with `Incident Originator`, which is a good descriptive name.
5. Select the **Incident** table.
6. Select **Contact** access as this permission is defined by a contact table relationship.
7. Select the table relationship created earlier in the *Creating contact lookups on the incident table* section in *Chapter 4.*
8. Configure all permissions to allow the originator to delete their own record.
9. Add both web roles that work in incidents – **Back Office** and **Foreman**:

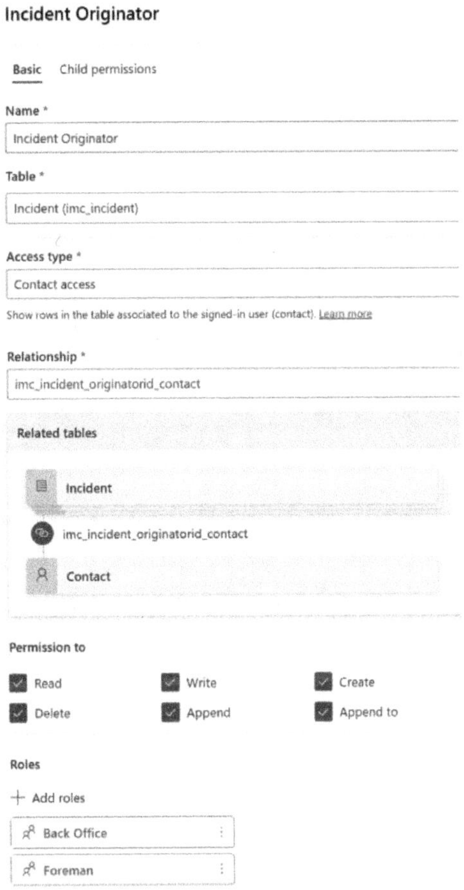

Incident Originator

Basic   Child permissions

Name *

Incident Originator

Table *

Incident (imc_incident)

Access type *

Contact access

Show rows in the table associated to the signed-in user (contact). Learn more

Relationship *

imc_incident_originatorid_contact

Related tables

■   Incident

⊕   imc_incident_originatorid_contact

A   Contact

Permission to

☑ Read   ☑ Write   ☑ Create
☑ Delete   ☑ Append   ☑ Append to

Roles

+ Add roles

ℛ Back Office   ⋮

ℛ Foreman   ⋮

Figure 5.4 – Contact access permissions to give the creator contact full access

Account access permissions

User contacts with **Back Office** and **Foreman** web roles in the organization need to have read access to all incidents reported under their account, as in all incidents within their organization. Sarah will achieve this with an **Account access** type.

Sarah ensured that all incidents were accessible to the logged-in user's account. To achieve this, she added a new field named **managing account** to the **Incident** table, designed as an **Account** lookup column. By configuring the **Account access** type to reference this **managing account** relationship, Sarah enabled all users associated with this account – essentially every contact linked to the organization – to access these incidents. This setup, shown in *Figure 5.5*, is applied across both the **Back Office** and **Foreman** web roles, guaranteeing access where needed:

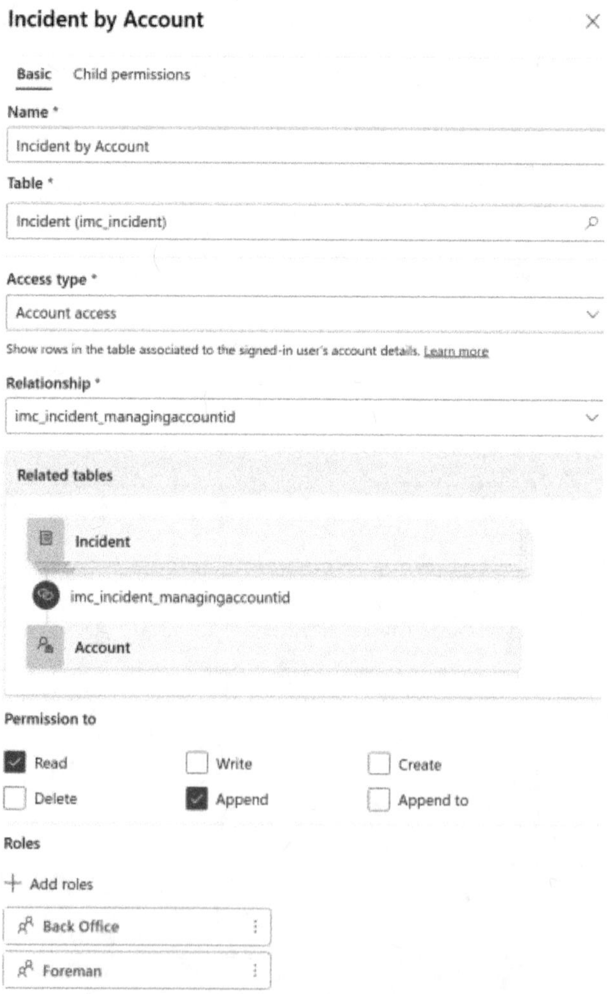

Figure 5.5 – Account access permissions to give the organization users read access

Child access permissions

In the context of our Incident Management feature, it's essential to manage the accessibility of notes associated with each incident. These notes often contain vital supplementary information, making their accessibility as crucial as the incidents themselves. To achieve this, Sarah implemented a **Child access** permission, which links the permissions of the notes directly to their corresponding incident records. This means that if a user has access to an incident, they automatically gain the same level of access to the notes attached to that incident. This linkage ensures consistency in data accessibility and simplifies permission management.

For example, if an incident record is accessible to a user under the **Account access** permission for the incident, the same user will have access to the notes under the **Child access** permission.

As shown in *Figure 5.6*, Sarah added a **Note** permission as a child to the **Incident** account permission as follows:

1. Open the Power Page studio.
2. Select the **Set up** tab, and then select **Table permissions**.
3. Select and open the account permission as created in the *Contact access permissions* section.
4. Select the **Child permissions** tab.
5. Select the **New** button, which opens the form as shown in *Figure 5.6*.
6. Configure the **Incident** table to **Note**, select the incident **Notes** relationship, and add the **Back Office** and **Foreman** web roles that have permission to incidents:

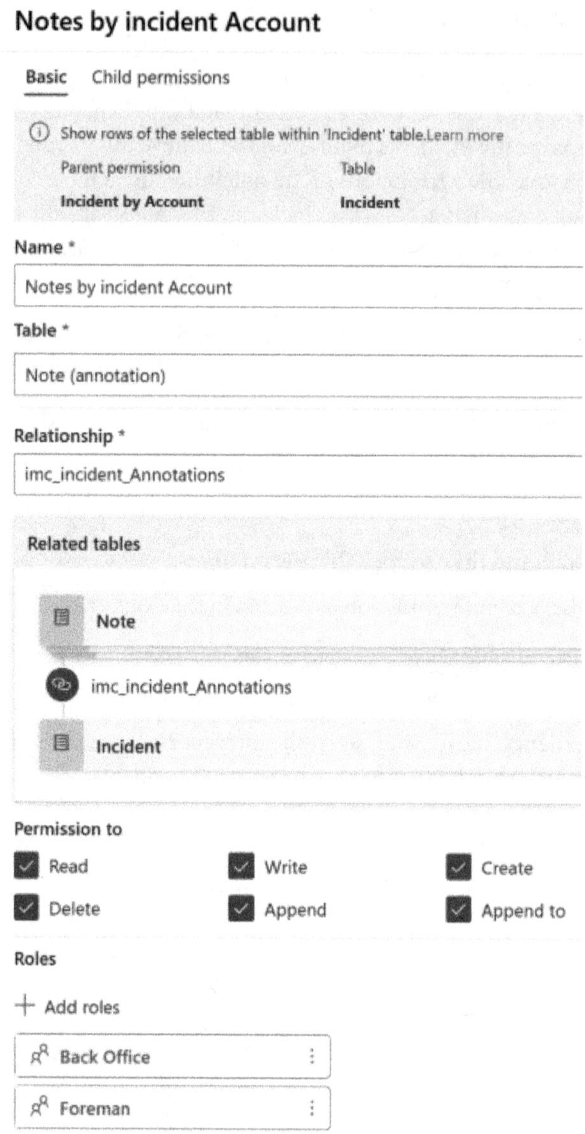

Notes by incident Account

Basic   Child permissions

ⓘ Show rows of the selected table within 'Incident' table. Learn more

Parent permission	Table
Incident by Account	**Incident**

Name *

Notes by incident Account

Table *

Note (annotation)

Relationship *

imc_incident_Annotations

Related tables

📇 Note

🔗 imc_incident_Annotations

📇 Incident

Permission to

☑ Read   ☑ Write   ☑ Create

☑ Delete   ☑ Append   ☑ Append to

Roles

➕ Add roles

⨝ Back Office	⋮
⨝ Foreman	⋮

Figure 5.6 – Child permissions

The architecture of table permissions

In this section, we introduce the architecture of table permissions within Power Pages and their integral role in securing Dataverse records. Understanding how to properly configure these permissions is crucial for maintaining data integrity and ensuring that only authorized users have access to sensitive information.

Table permissions in Power Pages are designed to fine-tune access control to Dataverse records. They are intertwined with web roles, which are assigned to users to define their access levels across the system. Here's how the architecture is laid out:

1. **Web roles**: These are sets of permissions assigned to users or groups of users. Roles determine what users can see and do within Power Pages. Web roles can be custom-created to fit the specific needs of different users, such as **Foreman** or **Back Office** roles in the context of the Rob the Builder company.

2. **Table permissions**: These permissions are assigned to tables within Dataverse to control access at the record level. Table permissions define what actions (**Read, Create, Write, Delete, Append, and Append to**) can be performed by users with specific web roles on specific tables. This granularity ensures that users only access data pertinent to their role within the organization.

3. **Configuration process**: These permissions unravel the access types that fortify data integrity with precision and finesse:

   - **Access types**: Table permissions are based on different access types that determine the visibility and modifiability of records:

     - **Global access**: Allows or restricts access to all records in a table, regardless of who created them. Ideal for tables containing non-sensitive, generic data.

     - **Contact access**: Restricts access to records created by or related to a specific contact. This is used when records have a direct relationship with specific users, such as incident records created by an employee.

     - **Account access**: Grants access to all records associated with an account. This might be used where employees of a specific company should see all incidents reported by their colleagues.

     - **Child access**: Used to manage permissions of records related to another record. For example, if a user has access to an incident, they automatically gain access to all notes attached to that incident.

4. **Linking table permissions to web roles**: Sarah configures table permissions and links them to appropriate web roles. This is crucial for ensuring that each role has the correct level of access, based on the organizational requirements and compliance standards. For instance, the **Foreman** role might have read and write access to incident records but only read access to customer contact information.

5. **Implementation example**: Sarah needs to ensure that foremen can view and update incident reports but cannot alter contract details. She sets up table permissions so that the **Incident** table allows read and write access for the **Foreman** role but restricts write access on the **Contact** table.

Practical applications and benefits

The structured approach to table permissions allows organizations such as Rob the Builder to do the following:

- **Secure sensitive data**: Ensuring that only authorized personnel can access or modify sensitive information

- **Comply with regulations**: Meeting industry standards and regulations by controlling who has access to specific types of data

- **Streamline operations**: By ensuring employees only have access to the data necessary for their roles, making navigation and usage of systems more straightforward and less prone to error

Privileges and relationships in Dataverse architecture

In Dataverse, privileges are the fundamental rights assigned to a role, defining what actions the role can perform across various tables. These privileges include **Create**, **Read**, **Write**, **Delete**, **Append**, and **Append to**, which can be finely tuned to align with business processes and security needs.

> **Role-based privileges**
>
> These are assigned at the role level and dictate the actions a user in that role can perform on any record within a table. For example, a role might have **Read** access to all customer records but **Write** access only to incidents they create.

Relationships

Dataverse uses relationships to link tables, which helps in organizing and securing data based on its relational context. Relationships can be one-to-one, one-to-many, or many-to-many, and these relationships help define how data access permissions are propagated.

> **Role-based relationships**
>
> These dictate how data in one table can be accessed based on data in another table. For instance, if there's a one-to-many relationship between the **Account** and **Incident** tables, a user with access to an account can be configured to access all related incident records, utilizing the **Account access** type.

Configuring access through relationships

Efficient management of access rights in relational databases hinges on the nuanced configuration of access through relationships. By establishing contextual controls and delineating child access permissions, organizations can sculpt finely tailored data security frameworks. Let's explore the pivotal mechanisms governing access control in relational data environments.

In the realm of access configuration, two pivotal strategies emerge to sculpt finely tailored data security frameworks – child access and contextual access controls:

- **Child access**: Permissions for a child record are directly linked to its parent. This is especially useful in hierarchical data structures where access to a parent record should grant access to related child records. For example, access to an incident might automatically grant access to all notes (child records) associated with that incident.

- **Contextual access controls**: Dataverse allows the setting of contextual permissions where access to records can be controlled based on related data. For example, an employee could be restricted to access only those records that are related to their department or projects they are directly involved in.

Dynamic row-level security

Dataverse implements **dynamic row-level security** (**DRLS**), which adjusts what data a user can see based on their assigned permissions and the relationships defined in the data model. This ensures that users only see data pertinent to their roles and responsibilities, thereby enhancing security and compliance.

Implementation of table permissions in Power Pages

When integrating with Power Pages, these table permissions and privileges are crucial for configuring sites that not only fetch and display data but also allow for secure data manipulation. For instance, in Sarah's case, configuring the **Incident** table with specific privileges linked to web roles ensures that foremen can update incident details directly through Power Pages, aligned with the permissions set in Dataverse.

By leveraging Dataverse's robust architecture, which supports complex relationships and granular privileges, Sarah can effectively manage who has access to what data within Rob the Builder's Power Pages application. This setup not only secures sensitive business data but also streamlines workflows by ensuring data is accessible to the right people when they need it, directly through the user-friendly interfaces of Power Pages. This approach lays a solid foundation for data integrity and secure user access within the Power Platform environment, making the IMS both efficient and compliant with stringent access and security policies.

Understanding the architecture of table permissions within Power Pages and their relationship with Dataverse is key for administrators such as Sarah. By setting up these permissions aligned with web roles, organizations can enhance security, comply with data governance policies, and ensure that their data ecosystems are both robust and user-friendly. As Sarah continues to configure these permissions, she ensures that the IMS is not only efficient but also adheres to strict access and security policies, setting a solid foundation for data integrity and secure user access within Power Pages.

Summary

This chapter focused on table permissions and security, and as Sarah progresses through configuring the various permissions, she's not only ensuring robust security for Rob the Builder's IMS. Each step brings her closer to developing a system that's not only efficient but also secure. The chapter explored how table permissions are crucial in controlling access to Dataverse records within the Incident Management feature. By configuring table permissions using the Power Pages studio, Sarah assigned specific permissions to web roles, enabling role-based security access. The chapter explained self-access permissions, granting users direct control over their own records. Additionally, Sarah addressed global access permissions, streamlining access to essential reference data. She also focused on contact access permissions, which restrict write and delete access to incident records to their creators. Child access permissions were another area Sarah mastered, ensuring uniform access rights across related tables. These steps in configuring table permissions have not only enhanced Sarah's understanding but also ensured a well-secured IMS for Rob the Builder.

By mastering these concepts, Power Pages users can effectively manage access and maintain data integrity within their applications.

Having now configured permissions and user access, in the next chapter, we will learn how to develop forms and web pages.

6

Basic Forms, Lists, and Web Pages

With the completion of Dataverse forms and table permissions in the previous two chapters, Sarah is ready to delve into the development of Power Pages forms and web pages. In this chapter, Sarah's focus will be on creating three essential forms: insert, edit, and read-only forms. Additionally, Sarah will design three corresponding web pages for each form, as well as a list page that provides convenient access to these forms as a hub or landing page.

Throughout this chapter, Sarah will utilize the Power Pages Management app to develop the Power Pages forms, enabling seamless integration and a cohesive user experience.

This chapter will cover the following topics:

- Basic forms
- Creating an insert form and web page
- Website sync – testing your work
- Creating an edit form and web page
- Creating an incident read-only page
- Creating a list page
- Creating an incident list page

Technical requirements

To progress through this chapter, you will need to have your Dataverse and Power Pages website provisioned, as shown in *Chapter 1*, and complete the development examples on **incident management (IM)** work of *Chapters 4* and *5*. In this chapter, we will use both the Power Pages studio and the Power Pages Management app, accessible from `https://make.powerpages.microsoft.com/`, as shown in previous chapters.

Integration of Dataverse tables with Power Pages forms, lists, and web pages

In Power Pages, the core functionalities of forms, lists, and web pages are deeply connected to Dataverse tables. These connections are essential for displaying, managing, and updating data dynamically on Power Pages. Here's how this integration works:

1. Forms and Dataverse tables:

    -   **Data binding**: Each form in Power Pages is directly linked to a Dataverse table. This binding allows the form to display data from the table and submit new or updated data back to it. For example, the **Incident insert** form Sarah creates is tied to the `Incident` table in Dataverse, enabling it to initiate new incident records.

    -   **Field mapping**: Fields on a Power Pages form correspond to columns in the Dataverse table. This mapping ensures that data entered into the form fields directly updates the appropriate columns in the Dataverse database.

2. Lists and Dataverse views:

    -   **Data retrieval**: Lists in Power Pages are configured to display views from Dataverse tables. These views are predefined queries that filter and sort data based on specific criteria, making it easier to display only relevant records on Power Pages.

    -   **Actionable insights**: Lists not only display data but also allow users to interact with it—such as editing, deleting, or viewing detailed information—directly influencing how data is managed in Dataverse.

3. Web pages and Dataverse data:

    -   **Dynamic content generation**: Web pages in Power Pages can dynamically display content from Dataverse tables using both forms and lists. This setup allows for real-time data updates, which is crucial for maintaining the accuracy and relevancy of information presented to the user.

    -   **Security and access control**: The integration ensures that data visibility and modifications are governed by the security roles and table permissions set up in Dataverse, which Sarah meticulously configures. This means that users see only the data they are authorized to access, maintaining data integrity and compliance.

This integration between Dataverse tables and Power Pages components ensures that Sarah can build robust, dynamic, and secure web applications efficiently. Understanding this connectivity is crucial as it lays the foundation for the practical applications discussed throughout this chapter.

This section has provided a clear and concise explanation of how Dataverse tables integrate with the various components of Power Pages, bridging technical explanations with practical implementation. This helps to reinforce the understanding of how data flows and is managed within the Power Pages environment, enhancing the overall comprehension of the system's architecture for the reader.

Basic forms

Basic forms in the Power Pages studio are data-driven configurations that allow end users to add forms to collect data on websites without requiring programming code. Basic forms are used to develop components for web pages and can support subgrids for building web applications. Basic forms are associated with web pages, and their properties control form initialization on the site. They require defining the target table, Dataverse form name, and mode (**Insert**, **Edit**, or **Read Only**). Configuration includes relationships to web pages, various settings for form behavior, and additional attributes such as geolocation and request validation.

> **Tip**
> For further reading on basic forms, see the following link: `https://learn.microsoft.com/en-us/power-pages/configure/basic-forms`

In the next sections, Sarah will learn to create and configure a basic form with **Insert** mode and its web page and to configure options and metadata properties for that form.

Creating an insert form and web page

First, Sarah will create an insert form and then set its options, add metadata to that form, and then create its corresponding web page. To implement this, Sarah will be using the Power Pages Management app to create and configure the form and the Power Pages studio editor to create and configure the web page.

An insert form is a basic form set in **Insert** mode. The form loads as a new record with no data and enables users to create a new record when the form is submitted.

To establish a user-friendly interface for the Incident Management feature, Sarah will use Power Pages Management for form creation and Power Pages Studio editor for web page setup. Key steps include the following:

1. **Creating an Incident insert basic form**: Setting up a form with the correct table and options in Power Pages Management

2. **Configuring basic form options**: Adjusting settings for user experience, such as tooltips and required fields

3. **Enabling attachments and associating table references**: Allowing file uploads and linking the managing account for access control

4. **Setting up basic form metadata**: Using metadata to preset field values such as incident date and originator

5. **Developing an Incident insert web page**: Creating and configuring a web page in the Power Pages studio, ensuring proper form selection and authorized access

6. **Finalizing web page settings**: Reviewing and adjusting page settings in Power Pages Management, including the title and template settings

These steps integrate the form into the website, providing a secure and efficient data entry interface. In the upcoming sections, Sarah will learn how to perform each of the preceding steps.

The Incident insert basic form

The **Incident insert** basic form is designed so that users can create a new incident and needs to be easy to use and guide the users in completing and submitting a new incident. Implement the form as follows:

1. Open Power Pages Management.

    I.   In the Power Pages studio, select the ellipsis (**…**) tab.

    II.   Select the **Power Pages Management** option.

2. In Power Pages Management, browse to **Basic Forms** under the **Content** tab. Select the **New** button, as shown in *Figure 6.1*.

3. This will open up a new basic form:

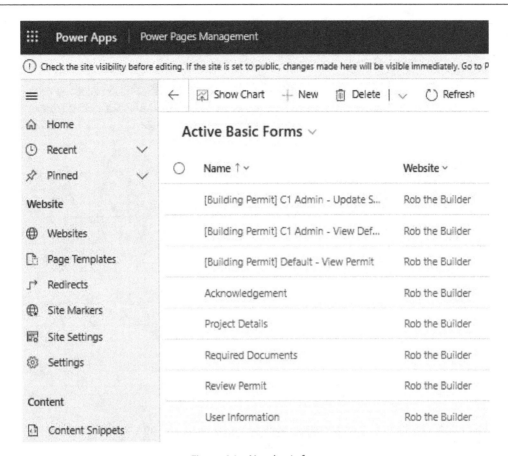

Figure 6.1 – New basic form

In the new basic form, Sarah will enter fields in the **General** tab, to enter the incident table and the Dataverse form details developed in *Chapter 4*.

Configure the **Incident insert** form shown in *Figure 6.2* as follows:

4. A meaningful name such as `Incident insert` makes it easier to group and work with many forms as the project develops.

5. Select **Incident** for the table name.

6. Select the **Portal** Dataverse form that we created in *Chapter 4*.

7. Select the **New Incident** Dataverse form tab that we created in *Chapter 4*.

8. Select the **Insert** mode and enable table permissions to apply table permission security to the form:

> **Tip**
> When selecting Dataverse forms, you can optionally select the **Dataverse** form tab. If you don't select a tab, the entire Dataverse form will render on the web page with the tabs in a vertical order.

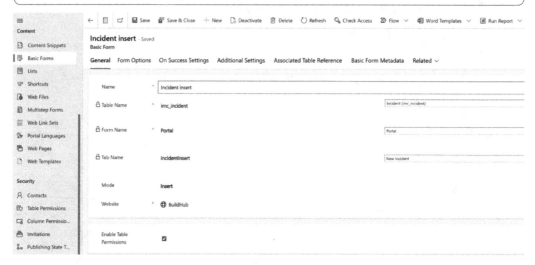

Figure 6.2 – Basic form General tab

Save the configuration, which creates a basic **Incident insert** form, and now Sarah can configure the form with options for improved user experience and data entry.

Form options

To effectively enhance user interaction and data accuracy on the website, Power Pages provides a range of form options. In the **Form Options** tab of the basic form, Sarah finds a suite of features tailored to refine the form's functionality and user experience. These options range from security measures such as CAPTCHAs to user guidance tools such as tooltips. While still in **Basic Form**, select the **Form Options** tab, as shown in *Figure 6.3*, and enter the following:

1. Check **ToolTip Enabled**.

2. Enable **Recommended Fields as Required**.

3. Check **Enable Validation Summary Links**:

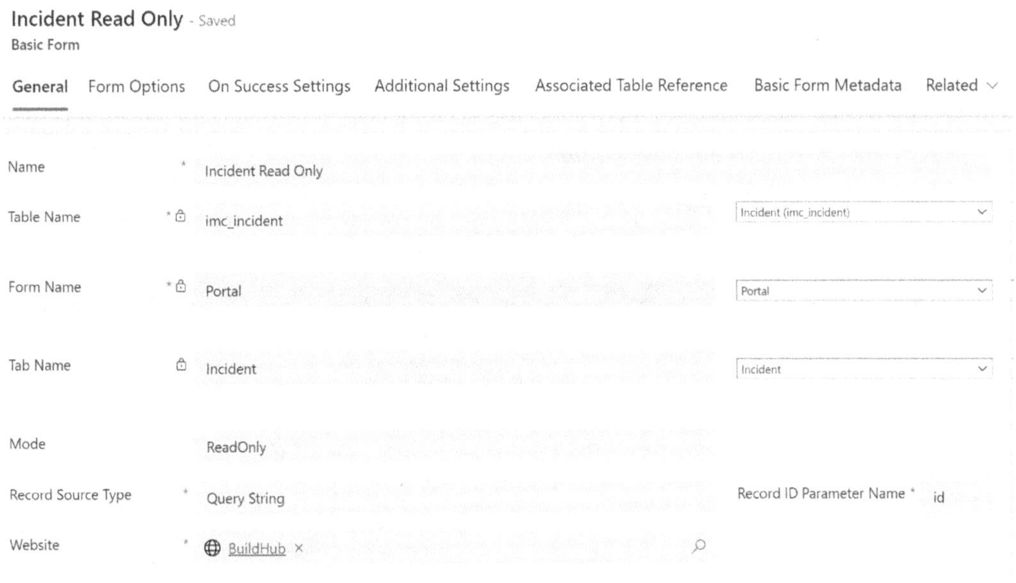

Incident Read Only - Saved
Basic Form

General   Form Options   On Success Settings   Additional Settings   Associated Table Reference   Basic Form Metadata   Related ∨

Name	*	Incident Read Only
Table Name	* 🔒	imc_incident	Incident (imc_incident) ∨
Form Name	* 🔒	Portal	Portal ∨
Tab Name	🔒	Incident	Incident ∨
Mode		ReadOnly
Record Source Type	*	Query String	Record ID Parameter Name * id
Website	*	🌐 BuildHub ×	🔍

Figure 6.3 – Basic Form options

Tooltips, sourced from a column's description, offer on-the-spot guidance to users, making data entry more intuitive. Additionally, Sarah enabled the **Set Recommended Fields as Required** option. It ensures users fill out all necessary information on the web page, thus maintaining data integrity and accuracy.

> **Tip**
> It is not practical to set many Dataverse columns as **Required**, but setting Dataverse columns as **Business recommended** and enabling **Set Recommended Fields as Required** makes them required on the web page.

Further fine-tuning of the form, including **On Success** settings, will be addressed once the **Incident edit** page configuration is complete, later in this chapter. Having completed our **Form Options** settings, let's see how Sarah will now set some additional settings.

Additional settings

In the **Additional Settings** tab, Sarah will focus on enabling file attachments, a feature implemented during the incident table creation in *Chapter 4*. This setting introduces a file upload control, enabling users to attach multiple files directly to their records. These attachments are seamlessly integrated as note attachments in the system. Here, you will find options not only for enabling file uploads but also for specifying storage locations, managing file types, and setting size limitations. This functionality is pivotal in enhancing data collection and providing a more interactive and comprehensive user experience.

As shown in *Figure 6.4*, configure attachments as follows:

1. Navigate to the **Additional Settings** tab within the **Basic Form** window.

2. Enable **Attach File**.

3. Select **Notes** for **Attach File Save Option**.

4. Enable **Allow Multiple Files**.

5. Select **Note Attachment** for **Attach File Storage Location**:

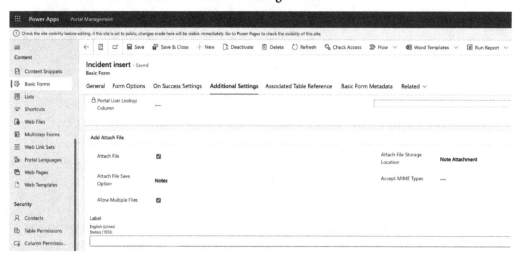

Figure 6.4 – Additional Settings configuration

By choosing not to specify anything for **Accept MIME Types**, Sarah ensures there are no restrictions on the file types that can be uploaded. Take time to look at the other options on this tab; these features show the sophistication and scope of Power Pages' codeless configuration.

Having completed the **Form Options** settings, Sarah needs to configure a data relationship reference being created with this record using the **Associated Table Reference** tab.

Associated Table Reference

In *Chapter 4*, while setting up the Incident table, Sarah introduced a lookup column linked to the account, labeled as **Managing Account**. This column plays a crucial role in controlling access to incident records. The objective is to ensure that only contacts associated with the managing account specified in the incident record can access these incidents. To implement this, Sarah configured the **Managing Account** lookup field to automatically populate based on the logged-in user's account. This process is executed in the **Associated Table Reference** tab of the form settings

Configure the **Managing Account** reference, shown in *Figure 6.5*, as follows:

1. Navigate to the **Associated Table Reference** tab within the **Basic Form** window.
2. Set the **Set table reference On Save** option to **Yes**.
3. Select **account** for **Table name**.
4. Select **Managing Account** for **Relationship Name**.
5. For **Source Type**, select **Record Associated to Current Portal User**.
6. Select **Contacts** for **Record Source Relationship Name**.
7. Enable **Populate Lookup Field**:

Figure 6.5 – Associated Table Reference tab

Configuring the **Managing Account** lookup column as an associated table reference to the current portal user, the person creating the incident record, will populate the **Managing Account** lookup column with the user's account; for example, their organization. Still working with data on the insert form, Sarah will now learn how to use metadata to create required information during the incident creation process.

Configuring basic form metadata

The **Basic Form Metadata** tab provides extensive customization options for modifying the behavior of not only form fields but also subgrids, tabs, and sections. This level of control allows you to enhance functionality and tailor the form's behavior to precisely meet your specific requirements. In this section, Sarah will explore the use of basic form metadata to set default values for fields. In our example, we can set the **Incident date** field to have a default value of **Now** and automatically populate the **Originator** field with the logged-in user. This functionality not only simplifies data entry but also serves practical purposes such as reporting and access rights, where the originator may have exclusive permissions to delete their own records.

> **Tip**
> For further reading, see the following *Microsoft Learn* link: `https://learn.microsoft.com/en-us/power-pages/configure/configure-basic-form-metadata`

To do this in the basic form, browse to the **Basic Form Metadata** tab and create a new metadata record, as shown in *Figure 6.6*:

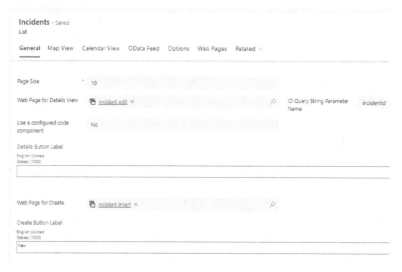

Figure 6.6 – New Basic Form Metadata record

This will open the **Metadata** entry form where a lot of features are available. Take some time to get familiar with some of these features. Here is where the codeless magic of Power Pages functions that would take lots of code to emulate in a website-coded project. In the next two examples, Sarah is implementing some simple automations; firstly, she needs to populate the originator column of the incident as the system needs to know who is creating the incident as they have elevated permissions and they are the reporter of the incident from an information point of view.

Configuring the Originator field

Select the **Attribute** originator and set the **Set Value On Save** option to **Logged-on user contact**. To set a lookup, you select the primary key ID, which for **Contact** is the `contactid` field, as shown in *Figure 6.8*. The **Originator** column does not need to be on the insert form, and when it is, it should be read-only:

Figure 6.7 – Metadata to select an attribute

There are many features on this long metadata configuration form; scroll down to the **Set Value On Save** section. By selecting **Current Portal User** as the **Type** value and selecting `contactid`, this is selecting the user who is creating (inserting) and submitting the incident record:

Figure 6.8 – Metadata to select a value for an attribute

This time, Sarah wants to configure a column on the form so that when the web page loads, it will have a value even before it is saved. The `Incident Date` column is for the originator of the incident to set the date and time of the incident.

Configuring the incident date to now

This field needs to be displayed on the form to allow the user to change it as it might have happened earlier. With the following configuration, we are setting the date and time to **Today's date** and want it prepopulated so that it has a value on the page load of the insert form:

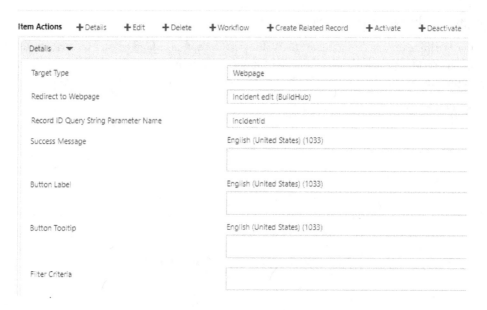

Figure 6.9 – Metadata to select a value for an attribute

Now, scroll down to the **Prepopulate Field** section so that Sarah can set the value of the column by selecting **Today's Date** for the **Type** option, which is the same as saying **Now** and will enter the date and time that the page loaded:

Figure 6.10 – Metadata to select Today's Date value for an attribute

Having learned how to implement the basic form configuration, in the next section, Sarah will create and configure the web page for this **Incident insert** basic form.

The Incident insert web page

The **Basic Form** configuration is best done in Power Pages Management, as most of the configuration features are not available for forms in the Power Pages studio editor; however, for proper settings, you have to use the Power Pages studio to create and configure web pages.

> **Tip**
> Further reading at *Microsoft Learn*: `https://learn.microsoft.com/en-us/training/modules/power-pages-studio/web-pages`

To create a new web page, follow the next steps:

1. Open the studio editor.

2. Select the **+ Page** button, as shown next. This opens the new page form, as shown in *Figure 6.11*:

Figure 6.11 – Power Pages studio: new page

3. Enter the page name as `Incident Insert`. Do not add it to the main navigation as this page's accessibility will be configured later in the chapter from a button.

4. For the layout, select the **Start from blank** option:

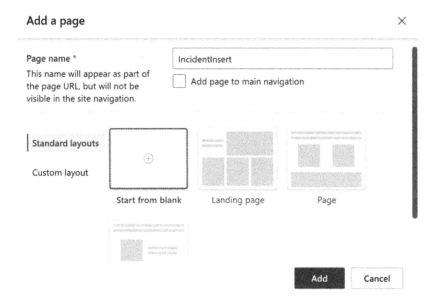

Figure 6.12 – Power Pages studio: new page form

Having created the new page, the form configuration is next. Sarah needs to configure the form to use on this page; for that, she selects the **Incident insert** form that she created earlier in this chapter, as shown in *Figure 6.13*:

Figure 6.13 – Selecting the Incident insert form

Sarah wants only authorized users with correct web roles to access this page, select the page permissions, and configure the page permissions as shown next. Here, Sarah will select the two web roles created in *Chapter 5*. This way, the page itself and its web link are hidden unless the logged-in user has one of these two web roles:

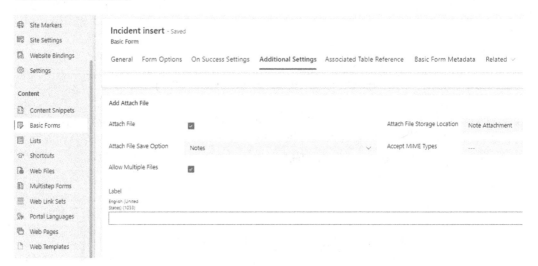

Figure 6.14 – Studio: Page settings

Now that the web page has been created, in the next section, Sarah will look at the detailed configuration of the web page in Power Pages Management.

Power Pages Management web page configuration

Sarah reviews the Power Pages Management page settings to check that the page has been configured correctly and to add a page title. The web template it has set is **Default studio template**. This web template is a default Power Pages web template and a good way to start development, even if later, Sarah will create and use her own custom web template:

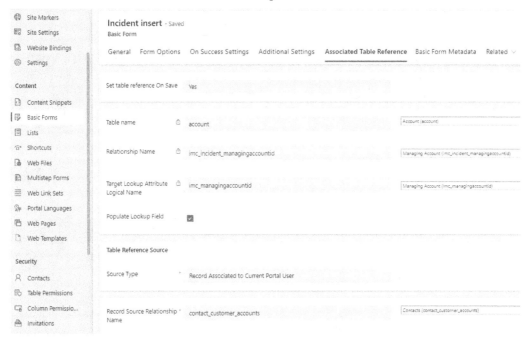

Figure 6.15 – Power Pages Management: root web page configuration

Creating a web page automatically creates two web page records: a root web page, as shown in *Figure 6.15*, and a content language web page, as shown in *Figure 6.16*. Both web page records have the same name. Sarah's website only has the **English** language installed and she only needs an English content page, which is created by default. If the project needed a multi-language site, then a content page for each language would be needed:

Figure 6.16 – Web page: Incident insert content page

As you can see in *Figure 6.16*, it has a **Title** field in which we will type `Incident` so that it appears as the title heading on the page and also displays the title as a prefix on the browser page window title, as shown next:

Figure 6.17 – Browser page window title

Having created the web page for the **Incident insert** form, we will want to test and review our work.

Website sync – testing your work

Power Pages are cached, and it can take up to 15 minutes for changes to appear. Pressing **Sync** updates changes to the cache immediately, so after making these changes in the Power Pages studio, press the **Sync** button found on the right-hand side for the changes to appear:

Figure 6.18 – Studio Sync button

After the sync, press **Preview** to view the page that is highlighted, which we have been working on, to see the following result:

Figure 6.19 – Incident insert page review

Having completed the **Incident insert** page, Sarah has learned the following:

- Sarah learned how to create and configure basic forms for insert functionality using Power Pages Management

- Knowledge of utilizing basic form metadata to modify the behavior of form fields allowing for enhanced customization

- The process of creating web pages using the Power Pages studio and configuring them with the appropriate forms

- How to set permissions for web pages, ensuring that only authorized users with specific web roles can access and interact with the pages

- The importance of syncing changes in the Power Pages studio to ensure that updates are immediately reflected in the cached web pages

Having completed the **Incident insert** form and its web page, in the next section, Sarah will learn how to create an **Incident edit** form and its web page.

Creating an edit form and web page

In the previous sections, Sarah successfully implemented a page for inserting new incidents, laying the groundwork for Rob the Builder's **incident management system (IMS)**. With the insert functionality in place, her next task is to enable users to modify existing incidents. This is a crucial feature, allowing for updates and additional information to be added as incidents evolve over time. To achieve this, Sarah now turns her attention to creating an **Incident edit** page. This page will not only allow users to view existing incident details but also make necessary changes, ensuring that the incident records are both dynamic and comprehensive. In this section, we will follow Sarah as she navigates through the process of creating and configuring an edit form, similar to what she did for the insert form, but with specific modifications to cater to the editing functionality.

The Incident edit basic form

In this section, Sarah will create and configure an edit form for the incident. As shown next, the configuration for the **Incident edit** basic form has an **Edit** mode, and Sarah selects the incident **Dataverse form** tab, created in *Chapter 4*. To configure the **Incident edit** basic form, Sarah creates a new basic form and configures the **General** tab fields, as shown in *Figure 6.20*, with the following steps:

1. Open Power Pages Management.

2. Select the **Basic Forms** tab and select the **New form** button.

3. Sarah selects **Edit** for the **Mode** setting, as shown in *Figure 6.20*.

4. Select **Query String** as the **Record Source Type** value:

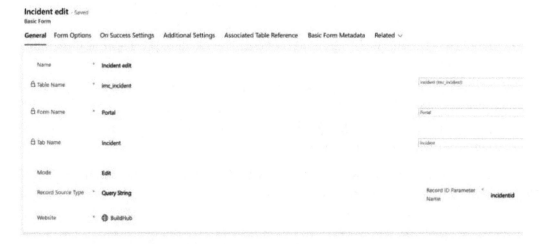

Figure 6.20 – Incident edit basic form

Note we have replaced the **Record ID Parameter Name** value with `incidentid` from the default of
`id`. It is useful to do this as you build more complex apps; sometimes, we need to build up a complex
query URL string that might contain several record IDs, and they will need unique parameter names.
Also, later in the book, we will show you how to use the query URL to develop dynamic breadcrumbs.
In both cases, the query string parameter name is used to reference records, and you need to distinguish
these query URL parameters by their name.

> **Tip**
> However, you can only change **Record ID Parameter Name** from `id` if the form is accessed
> as a web page and not as a (pop-up modal) form. **Record ID Parameter Name** must remain
> as `id` if access is by form and not web page.

Basic form options

As per how we implemented the **Incident insert** form, configure the form options. To do this while
still in the basic form, browse to the **Form Options** tab and enter the following fields, as shown in
Figure 6.21:

1. Check **ToolTip Enabled**.

2. Enable **Set Recommended Fields as Required**:

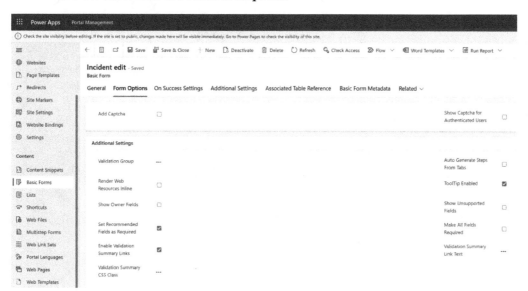

Figure 6.21 – Form Options tab

With the **Form Options** settings set, Sarah needs to allow attachments on the incident form so that
users can upload evidence and other documents of the incident.

Additional settings – Allowing attachments

Sarah needs to set attachments allowed with the configuration that enables the multiple file upload feature, as Sarah implemented in the previous section of this chapter:

1. In the basic form for **Incident edit**, select the **Additional Settings** tab.
2. Enable **Attach File**, shown in *Figure 6.22*.
3. Select **Notes** as the **Attach File Save Option** value.
4. Enable **Allow Multiple Files**.
5. Set **Attach File Storage Location** to **Note Attachment**:

Figure 6.22 – Configuring attachments

Having completed the basic form options for the **Incident edit** form, in the next section, Sarah needs to configure action buttons that allow users to run functions on an existing incident record.

Configuring action buttons

As an edit form, Sarah wants to give the user the ability to delete or deactivate incident records. Sarah does this by implementing action buttons with a **Delete** button and a **Deactivate** button with the following steps:

1. Browse to the **Additional Settings** tab, as shown in *Figure 6.23*.
2. Add a **Delete** action button.
3. Add a **Deactivate** button:

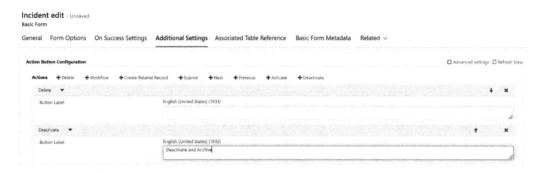

Figure 6.23 – Action buttons

4. Sarah changes the label of the **Deactivate** button to `Deactivate and Archive`.

5. On the **Delete** button, Sarah wants to add tooltips to the button and a confirmation message window, as shown in *Figure 6.24*.

6. Select the **Advanced Settings** tab, which shows all the available options on the action buttons.

7. Sarah adds a `Permanently delete this record` tooltip.

8. Configure the **Button Placement** setting to **Above Form**.

9. Configure **Button Alignment** to **Right**.

10. Sarah adds a confirmation message; by doing this, a confirmation window will pop up with that message before deleting the record:

Figure 6.24 – Incident edit form: action buttons' configuration

Having configured the form buttons, in the next section, Sarah delves further into the development of the IMS for Rob the Builder; her focus now shifts to the intricate details of form functionality.

Configuring Basic Form Metadata options

In this section, Sarah is set to configure **Basic Form Metadata** options for the **Incident edit** form. This process is essential as it allows her to enhance the form's usability and effectiveness, particularly in managing complex elements such as subgrids. By adjusting the metadata, Sarah aims to ensure that the subgrids not only display correctly but also provide users with intuitive and efficient action buttons for managing incident-related data. Additionally, she plans to optimize the user experience by transforming lookups into more user-friendly drop-down menus. These adjustments are key to creating a seamless and functional interface for the IM process, reflecting the dynamic and interactive nature of the system. Sarah will add metadata records to the **Incident edit** form to render the people subgrids that she placed on the form in *Chapter 5*.

Configuring subgrids

For subgrids to be viewable and have their own action buttons, Sarah needs to configure the subgrids as follows on the **Basic Form Metadata** page. Select the **Basic Form Metadata** page and add a new record. As shown earlier in the chapter, Sarah needs to browse to the **Basic Form Metadata** tab, add a **New Basic Form Metadata** page, select **Subgrid**, and then select the **PeopleInvolved** subgrid, as shown next:

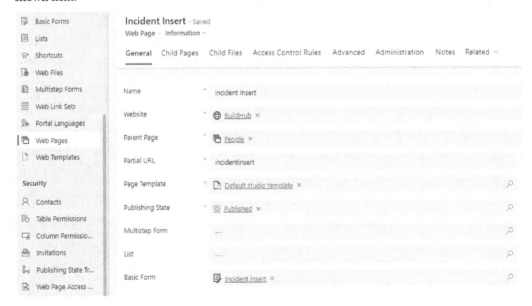

Figure 6.25 – Subgrid metadata Associate button configuration

Sarah also needs to add an **Associate** and **Disassociate** action button.

The **Associate** button will add people to the subgrid. As shown in *Figure 6.25*, to configure the **Associate** button, select the **Active Contacts** view to use for the **Lookup find** selection. Add an **Add Person Involved** button label and also other details such as a tooltip or success message. These subgrids display and represent tables in a many-to-many relationship, so to add a record, we associate, and to remove a record, we dissociate. Remember—these are contact records, and the records that are viewable are determined by the table relationships we created earlier in *Chapter 5*.

Now, configure the **Dissociate** button, as shown next; this will enable users to remove a person involved from the subgrid:

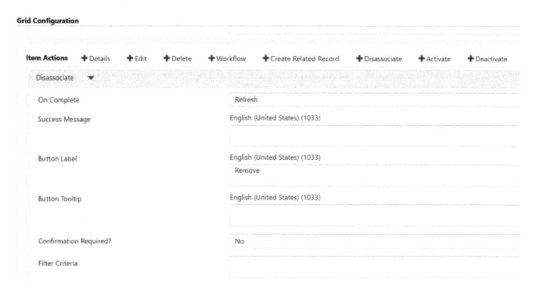

Figure 6.26 – Subgrid metadata Disassociate button configuration

Having completed configuring the **PeopleInvolved** subgrid, Sarah will configure the other two **PeopleInvolved** subgrids in a similar way, as shown in *Figure 6.27*:

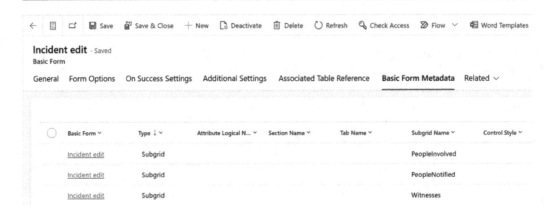

Figure 6.27 – Subgrid metadata three contact subgrids

Having completed the subgrids' configuration on the **Incident edit** form, Sarah needs to configure the notes control. It is a common requirement to add notes to a page so that users can add notes to a record—in this case, adding notes to an incident record. While the incident is active, Sarah needs to allow users to add notes to comments and provide evidence of the incident before it is completed.

> **Tip**
> Further reading on subgrids: `https://learn.microsoft.com/en-us/power-`
> `pages/configure/configure-basic-form-subgrid`

Notes control

Earlier in this chapter, Sarah added attachments in the incident forms in **Additional settings**, which enables users to upload attachments as notes. Previously, *Chapter 5* introduced the timeline control, which was added to the **Incident Dataverse** form. This enables the notes control on the form and leverages the notes functionality. Now, Sarah's next step involves configuring the notes control. She'll do this within a metadata record on the **Incident edit** basic form, enhancing the form's functionality. This enhancement enables users to add, delete, and update notes seamlessly. To achieve this, Sarah configures the notes control, as shown in *Figure 6.28*:

1. Open Power Pages Management.
2. Select the **Basic Forms** tab.
3. Select the **Incident edit** form.
4. Select the **Basic Form Metadata** tab.
5. Select the **+ New Basic Form Metadata** button.
6. Select a **Type** value of **Notes**, as shown in *Figure 6.28*.

7. Set **Create Enabled** as `true`.

8. Set **Edit Enabled** as `true`.

9. Set **File Attachment Location** as **Note Attachment**:

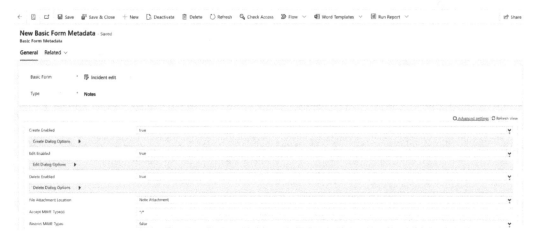

Figure 6.28 – Basic Form Metadata notes configuration

Metadata records also provide configuration to override or modify the behavior of fields, which, in the next section, Sarah will modify with a lookup control.

Configuring Incident Type as a dropdown

Sarah wants to configure the **Incident Type** lookup field as a dropdown as this will save user clicks, providing a better user experience, and for such a short list, it does not need a lookup search. Here's how she does that:

1. Select the **+ New Basic Form Metadata** button.

2. Select a **Type** value of **Attribute**, as shown in *Figure 6.29*.

3. Set **Control Style** as **Render Lookup as Dropdown**.

4. Set **Attribute Logical Name** as **Incident Type**:

Figure 6.29 – Basic Form Metadata Attribute control style configuration

Having completed the **Incident edit** basic form configuration, in the next section, Sarah will need to create an **Incident edit** web page, which uses the now completed **Incident edit** basic form.

The Incident edit web page

The **Incident edit** web page is used as part of the **Incident create** process and also for users to browse to open the incident record to review and edit the incident, adding notes and attachments as needed and adding people involved in the incident.

Sarah will configure the **Incident edit** web page as follows:

1. Open the Power Pages studio.
2. Create a new page, as shown earlier in this chapter.
3. For the layout, select the **Start from blank** option.
4. Select the **Incident edit** form we created in the previous section.

5. Configure the **Page settings Partial URL** setting with `incidentedit`, as shown in *Figure 6.30*:

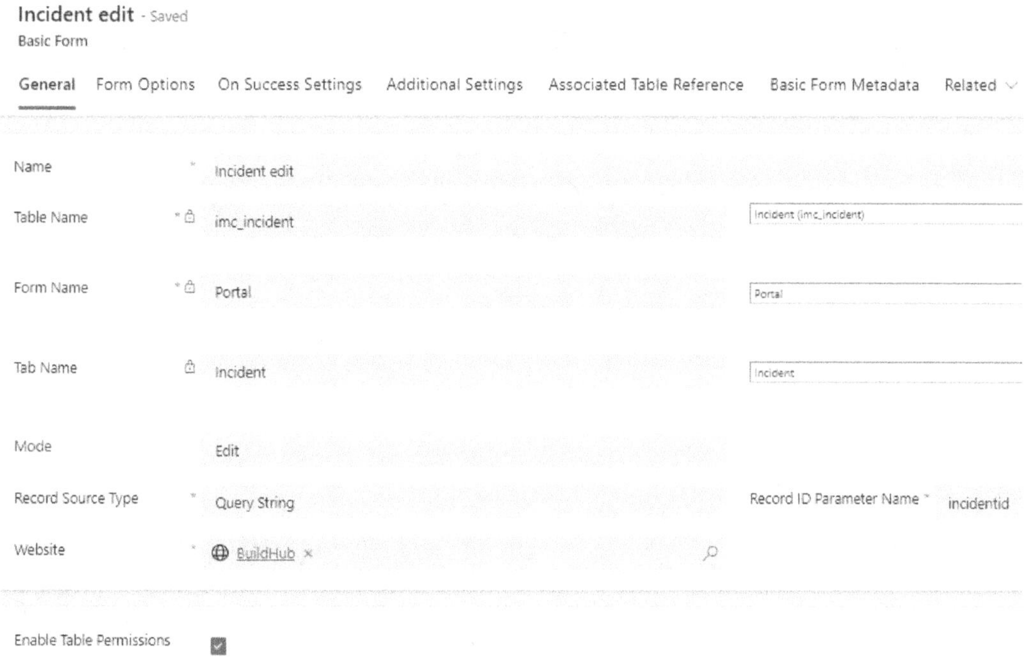

Incident edit - Saved
Basic Form

General   Form Options   On Success Settings   Additional Settings   Associated Table Reference   Basic Form Metadata   Related ⌄

Name   Incident edit

Table Name   * 🔒   imc_incident   Incident (imc_incident)

Form Name   * 🔒   Portal   Portal

Tab Name   🔒   Incident   Incident

Mode   Edit

Record Source Type   *   Query String   Record ID Parameter Name *   incidentid

Website   *   🌐 BuildHub ×   🔍

Enable Table Permissions   ☑

Figure 6.30 – Incident edit page setting

Having created the **Incident edit** web page, the content web page is automatically created for the default language, which for Sarah's website is **English**. Sarah browses to Power Pages Management app and selects the **Web Pages** tab. Here, Sarah opens and modifies the **Incident edit** content page so that she can add the title of `Incident` to the web page, as shown in *Figure 6.31*:

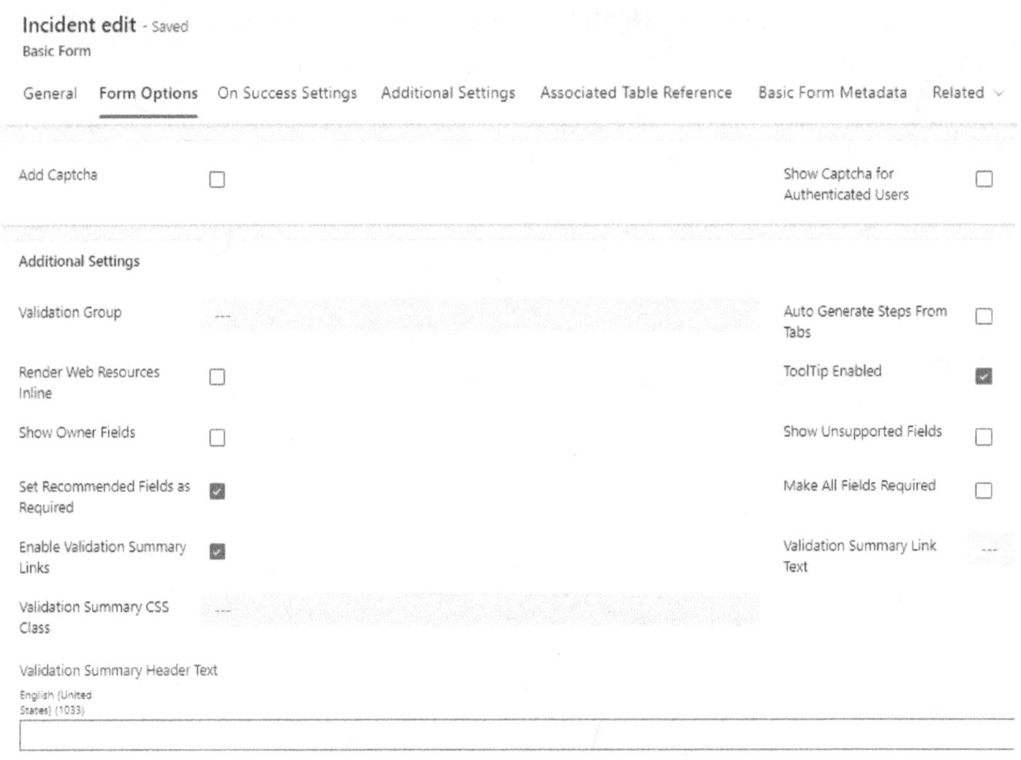

Incident edit - Saved
Basic Form

General   **Form Options**   On Success Settings   Additional Settings   Associated Table Reference   Basic Form Metadata   Related ∨

| Add Captcha | ☐ | | | Show Captcha for Authenticated Users | ☐ |

Additional Settings

Validation Group	---			Auto Generate Steps From Tabs	☐
Render Web Resources Inline	☐			ToolTip Enabled	☑
Show Owner Fields	☐			Show Unsupported Fields	☐
Set Recommended Fields as Required	☑			Make All Fields Required	☐
Enable Validation Summary Links	☑			Validation Summary Link Text	---
Validation Summary CSS Class	---				

Validation Summary Header Text
English (United States) (1033)

Figure 6.31 – Adding Incident as a title to the content web page

Now, having completed the **Incident edit** web page, in the next section, Sarah wants to configure an **Incident insert** page to redirect to this **Incident edit** page, giving the user a seamless experience. Sarah needs to do this so that the initial incident entry form is simple and then opens the relevant **Incident edit** form, which is configured to the **Type** value of **Incident**.

On Success Settings tab of the Incident insert form

The **On Success Settings** tab can be used to configure the behavior of the form when the user presses the **Submit** button.

In order to accommodate the completion of other form fields, Sarah wants to utilize both the insert and edit forms for an incident creation process, to implement a two-step design approach. This design strategy involves capturing the minimum required incident fields in the insert form and then redirecting the user to the edit page for further completion of the incident record. This two-step process is necessary because a crucial section of the incident form involves adding people in a subgrid. However, subgrids are only rendered in edit or read-only forms, not in insert forms. Hence, the redirection from the insert form to the edit form allows users to seamlessly finalize the incident by adding people to the subgrids.

Sarah will configure this design pattern as follows:

1. Browse to the **Incident insert** form.
2. Select the **On Success Settings** tab.

Here are the configurations to redirect to the **Incident edit** page, as shown in *Figure 6.32*:

1. On the web page, set the **Incident edit** page.
2. Enable **Append Record ID To Query String**, which on creation will add the new record ID to the URL.
3. Enter `incidentid` as the **Record ID Parameter Name** value:

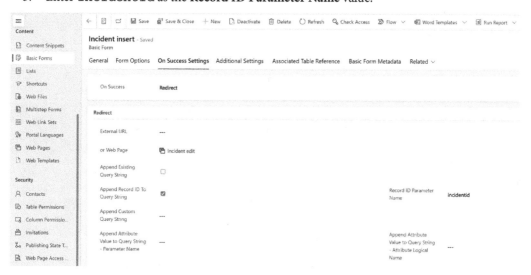

Figure 6.32 – Incident insert basic form on success configuration

Sarah has now completed the incident creation process, including **Incident insert** and **Incident edit** pages. In the next section, Sarah will create a read-only incident page; this is useful for rendering a form that is not editable.

Creating an incident read-only page

A read-only page is essential in Sarah's project for displaying incident details in a secure and uneditable format, ensuring that users can view critical information without the risk of accidental modifications or unauthorized changes.

Sarah will use the Power Pages studio to implement the read-only form version of the incident form and page as follows:

1. Open the Power Pages studio and add a new page, as shown in the previous section.

2. Enter a **Page name** value of Incident Read Only, as shown in *Figure 6.33*.

3. Disable **Add page to main navigation**.

4. Select layout as **Start from blank** as we will be using a form to fill the page's main content

5. Select the **Incident** table.

6. Select the **Portal** form, as shown in *Figure 6.34*:

Figure 6.33 – Incident Read Only page

Sarah is selecting the same **Incident Portal** form created in *Chapter 5* and used in all of the basic forms she created in this chapter. Sarah enters `Incident Read Only` as the new form name, as shown in *Figure 6.34*:

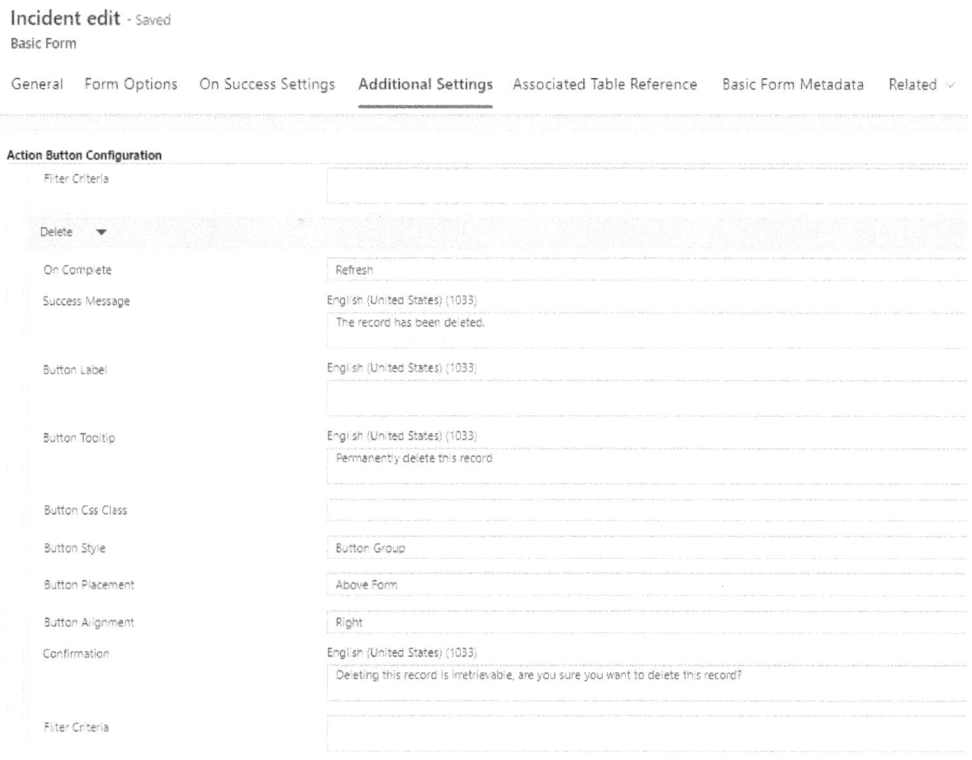

Incident edit - Saved
Basic Form

General   Form Options   On Success Settings   **Additional Settings**   Associated Table Reference   Basic Form Metadata   Related ⌄

Action Button Configuration

Filter Criteria

Delete   ▼

On Complete   Refresh

Success Message   English (United States) (1033)
                             The record has been deleted.

Button Label   English (United States) (1033)

Button Tooltip   English (United States) (1033)
                             Permanently delete this record

Button Css Class

Button Style   Button Group

Button Placement   Above Form

Button Alignment   Right

Confirmation   English (United States) (1033)
                             Deleting this record is irretrievable, are you sure you want to delete this record?

Filter Criteria

Figure 6.34 – Incident Read Only form

Sarah will open the Power Pages Management app to complete the form configuration, as most of the form configurations are only available in Power Pages Management. Sarah will configure the form mode as **ReadOnly**, as shown in *Figure 6.35*.

As a developer, it is most convenient to keep Power Pages Management open while working on a project. Sarah can also open the Power Pages Management app's basic form by clicking **Open Power Pages Management** in the Power Pages studio.

As shown in *Figure 6.35*, Sarah will set the form tab to **Incident**, which she created in *Chapter 5*; otherwise, the entire form with all tabs is rendered, which is not what Sarah wants in her page design. Sarah wants this page to display only the **Incidents Dataverse** form tab. The **Record ID Parameter Name** value is renamed by changing it to `incidentid` from `id` as we did with the edit page previously; this means that this form will only be displayed on a web page and not as a modal (pop-up) form, which would require keeping the **Record ID Parameter Name** value as `id`:

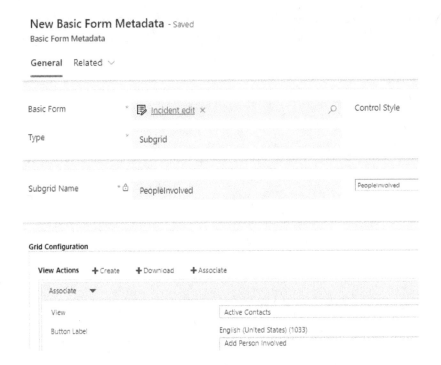

Figure 6.35 – Incident Read Only form configuration

In this chapter, so far, Sarah has learned to create and configure insert and edit forms and their web pages. Now, to complete the **CRUD** pattern (**Create, Read, Update, Delete**) to further enhance the Incident Management feature and provide an efficient way for users to navigate and manage incident records, Sarah will create a list page. This page will serve as a central hub or landing page, displaying all incident records in a dynamic and interactive format, allowing users to easily access, view, and perform actions on them as part of the complete CRUD pattern implementation.

Creating a list page

List page sites are dynamic grids that utilize the Dataverse table views we created earlier in *Chapter 5*, allowing a codeless way to display records and to have action buttons placed for CRUD operations on records displayed. It is most convenient to work in the Power Pages Management app to create and configure a list control and then to create its web page in Power Pages studio.

> **Tip**
> Further reading on lists: https://learn.microsoft.com/en-us/power-pages/getting-started/add-list

Sarah will create a list page as follows:

1. Open the Power Pages Management app.

2. Browse to the **Lists** tab and add a new list.

3. Enter the details as shown in *Figure 6.36*, entering `Incidents` for **Name** and selecting **Incidents** as the table.

4. Save the record. After saving, Sarah will add views and complete the configuration:

Figure 6.36 – Incidents list configuration with Incidents table

Now that the list is created, we can add the views we created earlier in *Chapter 5*, as shown next. This allows users to elect different views of lists, which could have different filter criteria, sorting, and column layouts for each view. Here, Sarah adds **All Active Incidents** and **Completed Incidents**:

Figure 6.37 – Adding views to a list control

Having created a list and added its views, Sarah wants to add action buttons, with which users can add a new incident or edit existing incidents on that list.

Configuring action buttons

Sarah will add a **Create** button to enable users to insert a new incident, which utilizes the **Incident insert** page we created earlier in this chapter. Sarah needs to configure the **Details** button to specify the **Incident edit** page she created earlier in this chapter; this creates a hyperlink to the first column on the list.

Sarah will implement this as shown in *Figure 6.38*:

- In the **List General** tab, for **Web Page for Details View**, select **Incident edit** page, as shown in *Figure 6.38*.

- Enter an **ID Query String Parameter Name** value of `incidentid`.

- For **Web Page for Create**, select the **Incident Insert** page:

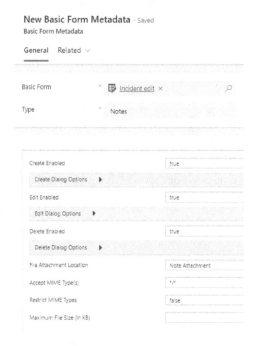

Figure 6.38 – List Create button and edit page configuration

List Search

Sarah wants to enable the **Search** option for the incidents list. Sarah will enable search, which is a full-text search of columns present in the view:

Figure 6.39 – Configuring a list search and enabling table permission

However, there is a comprehensive filtering feature also available, which is especially useful if you are providing a reporting list of large data.

> **Tip**
> Further reading on list filtering: `https://learn.microsoft.com/en-us/power-pages/configure/list-filter-configuration`

Most list pages need the ability to download. Power Pages downloads a `.csv` Excel file. To do this, Sarah will browse to the **Options** tab and add a **Download** button to the grid configuration, as shown next:

Figure 6.40 – List Download configuration

Sarah also wants to have **Details** as an action button to redirect to the edit page she created earlier in this chapter. On the **List Options** tab, click **Advanced Settings** and enter the following options to add a **Details** action button:

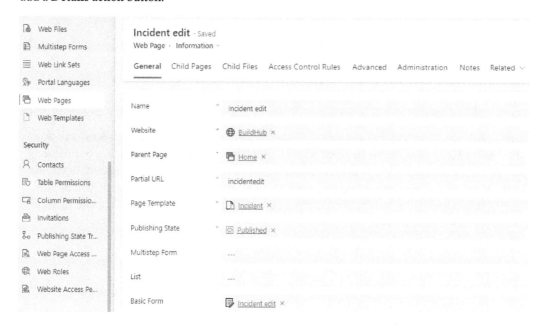

Figure 6.41 – List Details action button

Having completed the list configuration, Sarah now needs to create and configure its web page in the next section.

Creating an incident list page

Open the Power Pages studio and select **New page**, which will create a new web page in the list.

Name the page Incidents and select **Blank** as you will be choosing the incidents list we just created. Then, select the **Incident List** option in the **Studio Component** selector shown next:

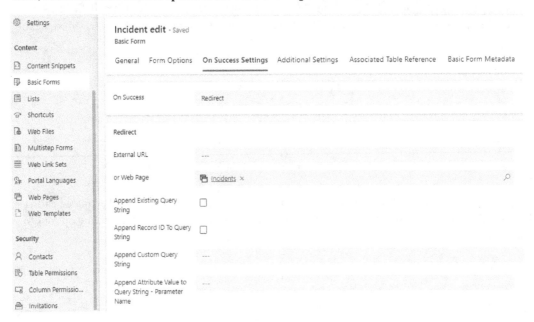

Figure 6.42 – Page component selector

Sarah only wants authorized users to be able to access this page or even see the web link to the web page. In the Power Pages studio, on the **Incidents** page, select **Page settings** and set the page permissions of the page to the two web roles of **Foreman** and **Back Office**, created earlier in *Chapter 2*. To test the work, Sarah needs to sync and preview the website and test the **Incidents** page, search, and buttons. Sarah will test that both the **New** button creates a new incident and **Details** opens the edit incident page and also test that **Search** filters results.

Summary

Chapter 6 served as a detailed guide on implementing the CRUD pattern in Power Pages, focusing on creating and configuring insert, edit, and read-only forms, along with their corresponding web pages. The chapter intricately detailed the process of form creation, starting from basic form setup to intricate configurations such as form options, associated table references, and metadata customization. These elements are crucial in developing functional and user-friendly forms that align with specific business requirements.

This chapter also emphasized the importance of web page management, where Sarah navigated through the Power Pages studio, setting up permissions and ensuring the synchronization of pages for accurate and secure user access. The chapter not only enhanced your understanding of Power Pages' capabilities but also demonstrated the practical application of these features in real-world scenarios. The focus was on imparting knowledge and skills to readers, equipping them with the tools to build robust, intuitive applications using Power Pages.

By the chapter's end, you gained valuable insights into the nuances of Power Pages, from form customization to managing web interactions. The chapter is a comprehensive resource for various skill levels, aiming to expand their proficiency in application development using Power Pages.

As we transition from the technical intricacies of Power Pages, the next chapter opens a new realm of possibilities with JavaScript and jQuery. Sarah's journey continues as she embarks on enhancing user interactions within forms, a task that beckons her to delve deeper into the dynamic world of scripting and web development. This next chapter promises to unfold the layers of JavaScript and jQuery, presenting them as vital tools for bringing interactivity and advanced functionality to Power Pages applications.

JavaScript and jQuery

Last week, Sarah delivered the incident management pages to Bob and Brenda for their review, and the initial feedback was positive. The pages performed exceptionally well on mobile devices, earning praise from the staff. However, a new set of challenges emerged as Brenda presented Sarah with a list of requested changes, mainly focused on improving user interactions within the forms. These changes required dynamic field visibility, as well as the ability to make fields mandatory or optional based on user input. Sarah knew that implementing these changes would require JavaScript, a skill she had yet to fully grasp. Thankfully, her experienced friend, Faisal, was ready to offer guidance. In this chapter, Sarah will explore the advice she received and the JavaScript and jQuery code she utilized to fulfill Brenda's requests.

Sarah will delve into the world of JavaScript and jQuery to implement dynamic form interactions. She will learn how to control field visibility and make fields required or optional based on user input. By showcasing practical examples and techniques, we aim to empower developers like Sarah with the skills needed to enhance their Power Pages projects.

JavaScript and jQuery are powerful tools within the arsenal of a Power Pages developer. Whether it's setting conditions to display certain fields, defining their parameters, or even implementing complex interactions, JavaScript and jQuery can significantly elevate the performance and user experience of websites.

In this chapter, Sarah will be introduced to these languages, giving her a solid foundation of their fundamental concepts, syntax, and application in Power Pages. Sarah will walk through examples of using JavaScript and jQuery to manipulate form fields, set up actions based on specific field values, and even manage the behavior of different data types in forms.

It's important to note that while having a background in JavaScript and jQuery can be helpful, it's not a prerequisite. This chapter is designed to cater to developers at all levels, ensuring everyone gains the insights and skills necessary to leverage JavaScript and jQuery effectively within Power Pages.

This chapter will cover the following topics:

- Understanding JavaScript in Power Pages
- Introduction to jQuery and its role in development
- The basic form JavaScript field
- The OnReady function
- Controlling field visibility with JavaScript and jQuery
- Managing field requirements with JavaScript and jQuery
- Understanding field syntax with JavaScript and jQuery
- Leveraging jQuery **Asynchronous JavaScript and XML (AJAX)** in Power Pages
- Using JavaScript and jQuery libraries

By the end of this chapter, Sarah will have a comprehensive understanding of how JavaScript and jQuery can be used to optimize her Power Pages. Additionally, the practical examples will provide her with hands-on experience, enabling her to implement these skills in real-world scenarios effectively.

JavaScript and jQuery – an overview

JavaScript is a high-level programming language that allows developers to enhance interactivity and perform complex operations on Power Pages. With JavaScript, Sarah can create dynamic updates and make changes, including but not limited to form field changes based on user actions, validations before submission, interacting with the data displayed on the page, and calls to the Dataverse web API.

jQuery and its role in Power Pages development

jQuery is a lightweight, feature-rich JavaScript library that simplifies tasks such as HTML document manipulation, event handling, and cross-browser compatibility. In Power Pages development, jQuery serves as a powerful tool for streamlining complex operations and reducing the need for extensive code. It fulfills the following roles:

- **Document Object Model (DOM) manipulation**: jQuery simplifies selecting and modifying HTML elements, allowing easy adjustments to be made to element visibility, content changes, and dynamic style modifications.
- **Event handling**: Developers leverage jQuery to attach event listeners to elements, enabling swift responses to user interactions such as button clicks and form submissions, resulting in interactive and responsive designs.

- **AJAX requests**: jQuery offers an efficient way to manage asynchronous data retrieval via AJAX requests. It empowers developers to fetch and present data from databases, external sources, or APIs without requiring a full page reload.

- **Animation**: jQuery boasts built-in animation functions that facilitate the creation of visually appealing transitions and effects. This enriches the user experience with elements such as sliding panels and fading images.

- **Plugin integration**: jQuery's extensive plugin library expands its capabilities, simplifying the incorporation of intricate features such as data visualization and form validation into Power Pages projects.

Having been introduced to JavaScript and jQuery and their applications, Sarah must learn where to place her code and where it is run in the **OnReady** function.

The basic form JavaScript field

In Power Pages, a JavaScript field exists on the basic form, the web from step and list, which serves as a critical space for executing custom JavaScript code. Developers use this field to implement functions such as data validation and actions that are triggered upon form submission. It's the designated area for embedding essential JavaScript functionality into Power Pages. Later in this chapter, Sarah will examine examples of how to add JavaScript code to make her changes; it's this field where she will place her new code.

Sarah can adapt JavaScript examples that are on the internet and contained in this book. In the next section, we will show you how and why she needs to wrap her code within an event called the OnReady function.

The OnReady function

The OnReady function is a jQuery event that runs as soon as the DOM is ready for JavaScript code to execute. This is important in Power Pages as it ensures that the JavaScript or jQuery code executes after the full web page is loaded, preventing any potential errors or mishaps.

Implementing the OnReady function

Sarah decides to create an example to manage fields in the incident edit basic form. The OnReady function event has the following syntax and function signature. To do this, Sarah must follow these steps:

1. Open Power Pages Studio.
2. Edit the website from the list of websites in this environment.
3. Select the **Edit Code** button. This will open Visual Studio Code in a browser window.
4. Browse to the **basic-forms** tab, as shown in *Figure 7.1*:

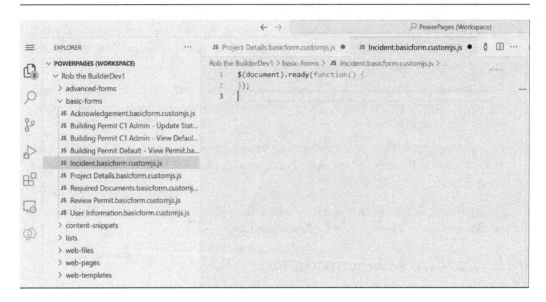

Figure 7.1 – Visual Studio Code basic-forms

5. Select the Incident basic form to edit its script.

6. Type in the following code:

```
$(document).ready(function() {
});
```

> **Tip**
>
> Developers should study the tabs and navigation for Visual Studio Code on the website. Here, developers can place any new code, including web templates, web pages, and content snippets.

Any code we place inside curly braces will run after the page loads.

Managing field requirements with JavaScript and jQuery

JavaScript and jQuery, with their ability to interact with the DOM and manipulate form fields, play a crucial role in achieving and managing form behavior functionality. In this section, we will examine some of the most common requirements for managing field visibility and how to set field requirements.

Controlling field visibility with JavaScript and jQuery

Manipulating field visibility is a common requirement in many applications. With JavaScript and jQuery, Sarah can show or hide fields based on a variety of conditions, such as the value of another field or user roles. In our example, we will add code to manage the visibility of fields in the incident edit basic form.

Showing/hiding a field based on a choice field value

Users have asked Sarah to show and hide injury fields to make it easier to guide the user through what fields need completing. We want the form to display the injury description field if there is any injury and hide the injury description field if there is no injury. In *Chapter 4*, we set this field as a checkbox, so it needs the `.is(":checked"));` syntax to evaluate if the checkbox is checked or not. We will create an `onchange` event to run when a choice field of any injury is changed. To implement this, enter the `AnyInjury` function code, as shown here:

```
function AnyInjury(){
    var typeVal = ($("#imc_anyinjuries_1").is(":checked"));
    if(typeVal){                    $("#imc_injurydescription").parent().
parent().show();
$("#imc_injurydescription_label").show();
    }else{                    $("#imc_injurydescription").parent().parent().
hide();
$("#imc_injurydescription_label").hide();
    }}
```

Place the preceding function underneath the `OnReady` function that Sarah, entered as shown in *Figure 7.2*:

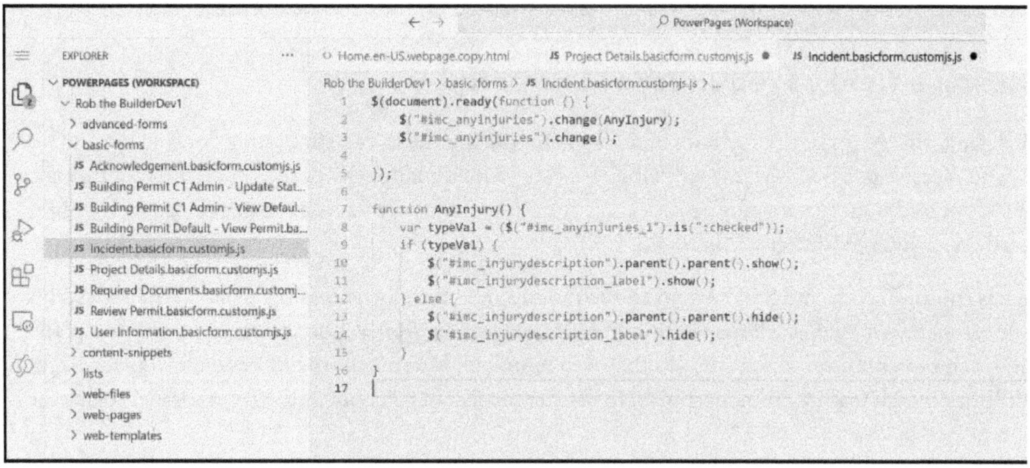

Figure 7.2 – Field event handler

Now, inside the `OnReady` function, we want to add the event handler code that will run every time the `Injury` field's value is changed:

```
$(document).ready(function() {
    $("#imc_anyinjuries").change(AnyInjury);
    $("#imc_anyinjuries").change();
});
```

The first line is the `onchange` event, while the second line, `$("#imc_anyinjuries").change();`, triggers the event so that the `AnyInjury` function runs on page load.

This `AnyInjury` function checks if the injury checkbox is checked and shows or hides the injury description fields accordingly.

Note that `var typeVal = ($("#imc_anyinjuries_1").is(":checked"));` checks if the checkbox with an ID of `#imc_anyinjuries_1` is checked.

If `typeVal` is `true` (the checkbox is checked), the injury description fields are shown.

If `typeVal` is `false` (the checkbox is not checked), the injury description fields are hidden.

This jQuery code's `OnReady` function ensures that the `AnyInjury` function is called when the document is ready and whenever the checkbox value changes.

`$(document).ready(function() {...});` ensures the code inside the function runs once the DOM is fully loaded.

`$("#imc_anyinjuries").change(AnyInjury);` sets up an event handler to call the `AnyInjury` function whenever the checkbox with an ID of `#imc_anyinjuries` changes.

`$("#imc_anyinjuries").change();` triggers the change event immediately on page load to ensure the initial state of the fields is set correctly based on the checkbox's current state.

Setting a field as required/not required

By default, the `required` attribute of a field is set during the initial configuration. However, there may be cases where Sarah wants to change this attribute dynamically based on certain conditions. JavaScript and jQuery provide the necessary functionality to set fields as required or not required based on various conditions.

This is common code and Sarah wants the required function to be accessible from all the JavaScript code across my website. As all the existing and planned web templates inherit **Page copy web template**, Sarah can put common functions into that web template. Microsoft provides code for Power Pages validators, which is what developers need to use for consistent behavior and error messages to appear for required fields.

> **Tip**
> To learn more about page validators, go to `https://learn.microsoft.com/en-us/power-pages/configure/add-custom-javascript#additional-client-side-field-validation`.

In this example, Sarah wants the injury description field to be required when any injury is selected and checked, thereby requiring the user to enter a description of the injury. Conversely, Sarah wants the field to be set as not required if there is no injury.

To do this, Sarah must follow these steps:

1. Open Visual Studio Code.

2. Browse to **Page copy web template** and add two functions, `SetFieldAsRequired` and `SetFieldAsNonRequired`, from the code listed in *Step 3*.

3. Since this is a web template, Sarah needs to wrap script tags around the code, as shown here:

```
<script type="text/javascript">
function SetFieldAsRequired(fieldName, displayname)
{
     if (typeof (Page_Validators) == 'undefined') return;
     if ($("#" + fieldName) != undefined && $("#" + fieldName +
"_label") != undefined)
     {
$("#" + fieldName).prop('required', true);
$("#" + fieldName).closest(".control").prev().
addClass("required");
          // Create a new validator object
          var Requiredvalidator = document.createElement('span');
          Requiredvalidator.style.display = "none";
          Requiredvalidator.id = fieldName + "Validator";
          Requiredvalidator.controltovalidate = fieldName;
var errormessage = "<a href='#" + fieldName + "_label'>"
+  displayName  + " is a required field.</a>";

Requiredvalidator.errormessage = errormessage;
          Requiredvalidator.initialvalue = "";
          Requiredvalidator.evaluationfunction = function ()
          {
               var fieldControl = $("#" + fieldName);
               if (fieldControl.is("span"))
               {
var value0 = $("#" + fieldName + "_0").prop("checked");
var value1 = $("#" + fieldName + "_1").prop("checked");
if (value0 == false && value1 == false)
                    {
                         return false;
                    }
                    else
                    {
```

```
                        return true;
                    }
                }
                else
                {
                    var value = $("#" + fieldName).val();
                    if (value == null || value == "")
                    {
                        return false;
                    }
                    else
                    {
                        return true;
                    }
                }
            };
    //Add the new validator to the page validators array
        Page_Validators.push(Requiredvalidator);
    }
}
function SetFieldAsNonRequired(fieldName)
{
    if (typeof (Page_Validators) == 'undefined') return;
    if ($("#" + fieldName) != undefined)
    {
        $("#" + fieldName).closest(".control").prev().
removeClass("required");
        $("#" + fieldName).prop('required', false);
        for (i = 0; i < Page_Validators.length; i++)
        {
if (Page_Validators[i].id == fieldName + "Validator")
            {
                Page_Validators.splice(i);
            }
        }
    }
}
</script>
```

Let's take a closer look at the SetFieldAsRequired function:

I.   **Check if Page_Validators is defined**: If not, exit the function.

II.   **Check if the field and its label exist**: If they do, proceed.

III. **Set the field as required**: Use `prop('required', true)` and add the `required` class to the closest control's previous sibling.

IV. **Create a new validator object**:

- Set `display` to none.

- Assign a unique `id` value for the validator.

- Set `controltovalidate` to the field name.

- Define an error message that links to the field label.

- Define `evaluationfunction` to check the field value.

V. **Add the validator to the Page_Validators array**: This ensures the custom validation is included in the page's validation process.

The `SetFieldAsNonRequired` function dynamically removes the required attribute from a field and removes its custom validator:

VI. **Check if Page_Validators is defined**: If not, exit the function.

VII. **Check if the field exists**: If it does, proceed.

VIII. **Remove the required class and attribute**: Use `removeClass('required')` and `prop('required', false)`.

IX. **Remove the validator from the Page_Validators array**: Iterate through the array and remove the validator with the matching `id` value.

The preceding code ensures that fields can be dynamically set as required or not required based on certain conditions, maintaining consistent validation behavior across the website.

At this point, the functions that set a field as required or not required are available from all the JavaScript on the website. Now, Sarah wants to implement this function in her incident edit form. To do so, she must add the function call to the `AnyInjury` function event handler she created earlier, as shown here:

```
function AnyInjury(){
var typeVal =($("#imc_anyinjuries_1").is(":checked"));
    if(typeVal){
        $("#imc_injurydescription").parent().parent().show();
$("#imc_injurydescription_label").show();
            SetFieldAsRequired("imc_injurydescription", "Injury
Description");
    }else{
        $("#imc_injurydescription").parent().parent().hide();
```

```
$("#imc_injurydescription_label").hide();
SetFieldAsNonRequired ("imc_injurydescription");
    }}
```

This function assigns the `injuries` Boolean field, which is rendered as a checkbox to the `typeVal` variable. Then, the code implements a condition if `typeVal` is true. If `typeVal` is true, then it shows the injury description field and sets the `injurydescription` field as required. Else (or otherwise) it sets the visibility of the field as hidden and sets the field as not required.

Understanding field syntax with JavaScript and jQuery

Power Pages use a different syntax for interacting with different types of fields in forms. Whether you're working with a text field, date field, choice field, or reference lookup field, understanding the correct syntax is essential for performing operations such as setting or retrieving values.

In the following examples, Sarah will explore the different syntax for a variety of field types and learn how to set or retrieve their values.

String field

Here is the syntax for retrieving the value for a `String` field:

```
var description = $("imc_description").val();
```

This is how to set a value for this field:

```
$("imc_description").val("Fire on site");
```

Number field

Here is the syntax for retrieving the value for a `Number` field:

```
var numberexample = $("new_numberfield").val();
```

This is how to set a value for this field:

```
$("new_numberfield ").val(10);
```

Date field

Here is the syntax for retrieving the value for a `Date` field:

```
var dateVal = $("#imc_date").val();
```

To set a value for a Date field add a suffix of `_datepicker_description` to the field name as in the incident date field:

```
$("#imc_date_datepicker_description").val("15/02/2023");
```

Choice field

Use the following code to retrieve the choice value. This is the number value of the choice, not its display name:

```
var optionsetvalueexample = $("#imc_response").val();
```

Set a value for the choice by setting the choice number's value:

```
$("# imc_response ").val(176230001);
```

Choice field with checkboxes

To retrieve a Boolean value of true or false from a two-choice checkbox, add `_1` as a suffix to the field name:

```
var anyInjuriesVal =  $("#imc_anyinjuries _1").is(":checked");
```

Pass in a Boolean variable, like so:

```
$(" #imc_ anyinjuries _1").prop('checked', anyInjuriesVal);
```

To set the Boolean choice field rendered as a checkbox as `true`, run the following code:

```
$(" #imc_ anyinjuries _1").prop('checked', true);
```

You can do the same for `false`:

```
$(" #imc_ anyinjuries _1").prop('checked', false);
```

Reference lookup field

To retrieve the lookup field value, return the GUID of that lookup:

```
Var originatorid = $("#imc_ originatorid").val();
```

A lookup field has three properties – `id`, `name`, and `entityname` – so we must set these three properties. We might retrieve these properties in Liquid statements from the query URL, at which point, we must populate the lookup field on page load to automatically fill a lookup field. We will take a closer look at this in the next chapter:

```
$("#imc_ originatorid").val(originatorid);
$("#imc_ originatorid_name").val("Hala Hassouna");
$("#imc_ originatorid_entityname").val("contact");
```

With that, Sarah has learned how to manage fields and retrieve and set values on a page using JavaScript.

The preceding code can be placed in a basic form in JavaScript if it's only relevant to the basic form and the code to be run on the form. The Cide Place Dina web template is common to all pages that use that web template, such as Page copy, which is common to all of Sarah's pages as it is a base web template for all her web templates. Code can also be placed on a web page. In this case, it will run across the web page and load before the basic form has loaded.

Now, Sarah needs to learn how to use AJAX to further enhance her pages. Using AJAX may seem complicated and advanced but many examples on the internet show how to implement an AJAX control.

Leveraging jQuery AJAX in Power Pages

jQuery's AJAX capabilities allow Sarah to perform asynchronous HTTP requests that can be leveraged in Power Pages to create dynamic, responsive applications. Low-code developers like Sarah are going to be asked to include controls such as charts, sliders, and carousels, so gaining some basic skills to implement AJAX controls will be useful.

> **AJAX introduction**
>
> A short introduction to AJAX can be found at `https://www.geeksforgeeks.org/ajax-introduction/`.

Let's take a look at a simple example of how AJAX can be implemented. Consider a situation where Sarah has a form and she wants to fill in a dropdown based on the user's selection in another dropdown. Sarah can use jQuery's AJAX method to get the data from the server and fill the dropdown without reloading the page:

```
$("#firstDropdown").change(function() {
    var selectedValue = $(this).val();
    $.ajax({
        url: "/path/to/api/endpoint",
        data: {
            selectedValue: selectedValue
        },
```

```
        success: function(response) {
            var secondDropdown = $("#secondDropdown");
            secondDropdown.empty();
            $.each(response, function(index, item)
{
                secondDropdown.append($("<option></option>").
val(item.value).text(item.text));
            });
        }
    });
});
```

This example code is designed to respond to a change in the `firstDropdown` element's value. When the user makes a selection in this dropdown, an AJAX request is triggered to a server-side API specified by the `url` parameter. The value that's selected from the first dropdown is sent as data, and upon a successful response from the server, `secondDropdown` is cleared and populated with options that have been retrieved from the API's response, dynamically updating the dropdown based on user input.

With this example, you now know where to put code and examples that you wish to use to enhance the user experience. Inevitably, Sarah encounters bugs and errors where pages don't load due to code errors. In the next section, we will introduce debugging.

Debugging JavaScript and jQuery

Debugging is an essential aspect of development. For JavaScript and jQuery in Power Pages, browser developer tools are usually the best place to start.

For example, Sarah can use the **Console** area in Chrome's **Developer tools** to execute JavaScript code on the fly and print output for debugging. From here, you can test and run JavaScript code in your web browser. For example, if Sarah wants to check if her JavaScript code is working as expected, she can open the **Console** area, enter her code, and see the results instantly. It's like having a whiteboard where she can scribble JavaScript ideas, and the browser will respond with what happens when she runs it.

Here's a simple example: Sarah wants to verify that her Boolean code for injuries is working. She opens the **Console** area, types `$("#imc_anyinjuries _1").is(":checked")`, and hits *Enter*. The **Console** area will instantly display the Boolean value for any injuries field, helping her verify that the JavaScript code is working correctly.

JavaScript's `console.log()` function allows her to output any variable or other data to the console, which can be very useful for tracking the flow of the code and the state of variables at various points.

For more advanced debugging, Sarah can use the **Sources** tab in Chrome's **Developer tools** to set breakpoints in the code and step through it one line at a time. This can help her see exactly what's going on at each step of the program.

Finally, a commonly used tool for debugging is the `alert` function. This will display an alert window with any values entered, such as `alert($("#imc_anyinjuries_1").is(":checked"));`, which will display the value of the Boolean field in an alert window.

> **Tip**
> Microsoft Learn has provided a helpful guide on debugging on the Edge browser: `https://learn.microsoft.com/en-us/microsoft-edge/devtools-guide-chromium/javascript/`.

Using JavaScript and jQuery libraries

JavaScript and jQuery libraries can significantly extend the functionality of Power Pages, from improving user interfaces to adding advanced features.

To include an external library in a website, developers typically add a `<script>` tag to the HTML that points to the library's URL. Developers can add this tag to the page or form. For example, suppose Sarah wants to use a jQuery UI library. Sarah would add a reference to it in the form like this to add the Chart library:

```
<!-- Include Chart.js library -->
<script src="https://cdn.jsdelivr.net/npm/chart.js"></script>
```

Sarah can then use the features provided by the jQuery UI library on her website. The `Chart.js` library is a powerful, flexible, open-source charting library for designers and developers. Sarah can use it to add interactive charts to her Power Pages.

Using Chart.js in Power Pages

Sarah wants to create a pie chart to represent the distribution of incidents by their status:

1. First, prepare the data. For this example, let's assume Sarah retrieved the incident data from Power Pages and it looks something like this:

```
var incidentsData = [
  { 'status': 'Open', 'count': 10 },
  { 'status': 'Completed, 'count': 15 },
  { 'status': 'Cancelled, 'count': 2 }
];
```

2. Next, create the arrays that will hold our labels (incident statuses) and data (incident counts):

```
// Extract labels and data from incidentsData
var labels = incidentsData.map(function(e) {
  return e.status;
});
var data = incidentsData.map(function(e) {
  return e.count;
});
```

3. Then, create a new chart:

```
// Create a pie chart
var ctx = document.getElementById('myChart').getContext('2d');
new Chart(ctx, {
  type: 'pie',
  data: {
    labels: labels,
    datasets: [{
      data: data,
      backgroundColor: ['rgba(255, 99, 132, 0.2)', 'rgba(54,
162, 235, 0.2)', 'rgba(255, 206, 86, 0.2)', 'rgba(75, 192, 192,
0.2)'],
      borderColor: ['rgba(255, 99, 132, 1)', 'rgba(54, 162, 235,
1)', 'rgba(255, 206, 86, 1)', 'rgba(75, 192, 192, 1)'],
      borderWidth: 1
    }]
  }
});
```

This code snippet prepares and displays a pie chart based on incident data. It starts by organizing the incident data into labels (incident statuses) and data (incident counts). Then, it utilizes the Chart.js library to create a pie chart using the prepared data. The resulting chart visually represents the distribution of incident statuses, allowing for data such as open, completed, and canceled incidents to be visualized with ease.

For this example, Sarah has used static data. In the next chapter, Sarah will use Liquid to implement a chart with dynamic data fetched from Dataverse tables.

Summary

In this chapter, Sarah's Power Pages journey took a deeper dive into the realms of JavaScript and jQuery. Her exploration began with understanding the critical role these scripting languages play in enhancing website interactivity and complex operations. She mastered the OnReady function and the basic form JavaScript field, both of which are integral components for successful JavaScript execution.

Sarah's foray into manipulating field visibility and dynamic requirement settings allowed her to cater to complex form behaviors. She became adept at handling various field types, employing JavaScript and jQuery to set and retrieve values effectively. Her journey included harnessing the power of jQuery AJAX for responsive application design, an invaluable skill in modern web development.

Furthermore, Sarah's venture into debugging and integrating external libraries underscored the importance of troubleshooting and expanding functionality beyond basic templates. The implementation of the Chart.js library marked a significant milestone, enabling her to visually represent data, thereby enhancing user engagement and understanding.

Overall, this chapter marked a pivotal point in Sarah's development journey, equipping her with essential skills and knowledge to optimize Power Pages using JavaScript and jQuery. These insights not only bolstered her technical prowess but also prepared her for upcoming challenges in dynamic web development. By understanding and applying these concepts, Sarah can now leverage JavaScript and jQuery in real-world scenarios, transforming the performance and user experience of her Power Pages.

As Sarah's journey continues in the next chapter, she'll dive into the Liquid language, marrying it with her newfound JavaScript skills. She will discover how to retrieve server-side data effortlessly and use Liquid in conjunction with JavaScript to create impactful and efficient code blocks. This exploration is set to further refine Sarah's ability to craft dynamic and user-centered Power Pages applications.

8

Web Templates and Liquid

In this chapter, Sarah uncovers the dynamic duo of **web templates** and **Liquid**. Web templates are a powerful feature of Power Pages and allow Sarah to define reusable, dynamic layouts. Liquid is a flexible templating language used for processing and outputting data using templates.

Sarah understands that she will need a web template for incidents so that she can provide customizations to web pages to implement conditional behavior in forms and retrieve related data when loading the page to provide a better user experience.

By utilizing web templates and Liquid in conjunction, Sarah creates dynamic pages that fetch and display data from Dataverse tables, interact with forms, and enable more advanced functionality in Power Pages. Whether it's filtering and sorting data, formatting outputs, or executing conditional logic, you'll find that these two tools give you a vast degree of control over the look, feel, and functionality of your Power Pages.

Regardless of her current skill level, this chapter helps Sarah understand and leverage the combination of web templates and Liquid to create more engaging and interactive Power Pages experiences. Sarah finds it helpful to start by copying and saving from an existing web template, making modifications as needed but still using the basic form.

This chapter covers the following topics:

- Web templates overview
- Creating a web template
- Introduction to Liquid
- Getting started with Liquid in web templates
- Using Liquid objects in Power Pages
- Using Liquid tags in Power Pages
- Using Liquid filters in Power Pages

- Advanced Liquid concepts for Power Pages

- Debugging web templates and Liquid

- Best practices for using web templates and Liquid in Power Pages

Web templates overview

Web templates provide a structured framework for creating consistent web pages. Think of web templates as the blueprints that guide the design and functionality of your pages. They serve as a foundation upon which you can build a cohesive user experience and ensure a unified look and feel across your entire portal.

By using web templates, you can streamline the development process and maintain consistency throughout your portal. They enable you to establish a standardized structure and design, making it easier to manage and update your pages. Additionally, web templates allow for the efficient reuse of common elements, reducing the need for redundant coding and ensuring a seamless user experience across different sections of your portal. Web templates contain the following:

1. Liquid

2. HTML

3. JavaScript and jQuery

Common Code

Web templates can include additional web templates, so use common code in the included web template; for this, the author likes to use the **page copy web template**. JavaScript, HTML, or Liquid code in the page copy web template is then available to run in all web templates that include the page copy web template. An example of the default studio web template, as seen using the Visual Studio code, is shown in *Figure 8.1*:

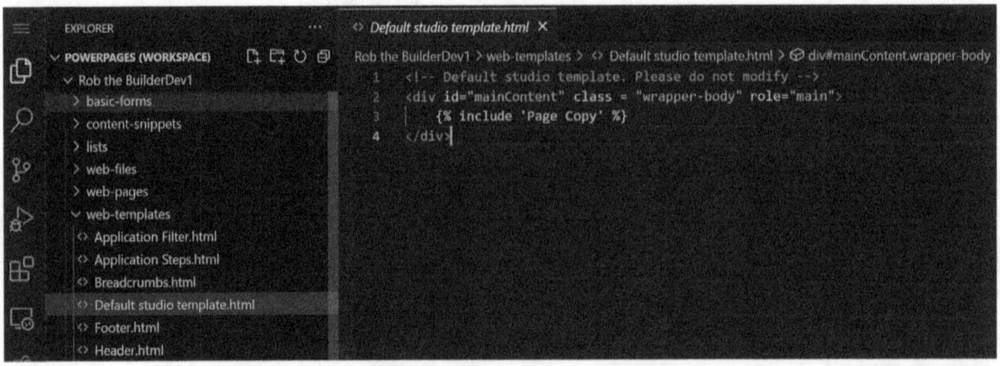

Figure 8.1 – Access web templates in Visual Studio code

In the following sections, Sarah will delve into the details of creating, customizing, and leveraging web templates in Power Pages.

Web templates example

Sarah examined the web templates installed in new websites. Sarah took a closer look at the default studio template, which all new pages are set with.

Here is the default studio template:

```
<!-- Default studio template. Please do not modify -->
<div id="mainContent" class="wrapper-body" role="main">
    <div class="page-copy">
       {% editable page 'adx_copy' type: 'html', liquid: true %}
    </div>
</div>
```

This default template provides a basic structure for web pages within Power Pages. Here's a breakdown of its components:

- **Container**: The `<div id="mainContent" class="wrapper-body" role="main">` element serves as the main container for the page's content. The `"wrapper-body"` class is typically used for styling purposes, and the `role="main"` attribute helps with accessibility by indicating the main content of the page.

- **Page copy**: Inside the main container, there is another `<div class="page-copy">` element that acts as a section for the main content of the page. This is where the bulk of the page's content will be displayed.

- **Editable region**: The `{% editable page 'adx_copy' type: 'html', liquid: true %}` tag is a Liquid template tag that makes the content within this region editable from the Power Pages studio. `type: 'html'` indicates that the content will be HTML and `liquid: true` allows the use of Liquid templating language within this editable region.

> **Important note**
>
> For more info about web templates, please review this link: `https://learn.microsoft.com/en-us/power-pages/configure/web-templates`.

In keeping with Sarah's wish to attain a professional approach, Sarah prepares a use case or Agile user story to implement web templates in her project.

Use case – Agile user story

Title: Creating dynamic web pages with web templates and Liquid

As a web developer

I want to create dynamic web pages using web templates and Liquid so that the platform can efficiently display data from Dataverse, provide a consistent user experience, and help implement conditional logic and data validation.

Acceptance criteria

1. The web template should define a reusable layout for web pages.
2. The web template should include dynamic elements using Liquid to fetch and display data from Dataverse.
3. The page should provide conditional behavior to show or hide sections based on user roles and data conditions.
4. The web template should support form interactions, including data validation and dynamic field requirements.

Sarah understands that she will need web templates to provide customizations to the web pages to implement conditional behavior on the forms and retrieve related data when loading the page to provide a better user experience.

By utilizing web templates and Liquid in conjunction, Sarah creates dynamic pages that fetch and display data from Dataverse tables, interact with forms, and enable more advanced functionality on Power Pages. Whether it's filtering and sorting data, formatting outputs, or executing conditional logic, you'll find that these two tools give you a vast degree of control over the look, feel, and functionality of your Power Pages.

Creating a web template

A web template serves as the foundation for structuring and designing your web pages in Power Pages. It provides a blueprint that defines the layout, structure, and functionality of a web page, ensuring consistency and efficiency across your portal. In this section, Sarah learns the process of creating a basic web template using Power Pages.

Accessing the web template editor

To begin creating a web template, navigate to the **Power Pages Management** app. From the app's navigation menu, locate and click on **Web Templates**. This will open the web template editor, where Sarah can define the structure and content of the web template. In this example, Sarah will create an **Incident Web Template** and inherit the **Page Copy** template that contains common code and the **Page Header** template, as shown in *Figure 8.2*:

Figure 8.2 – Incident web template

This web template for incidents, written in HTML with Liquid templating language, structures a web page into specific sections:

- **Container**: The entire template is enclosed within div with the container class, creating a central container for the web page's content.

- **Page heading**: Inside the container, div with the page-heading class serves as a section for the page's heading elements.

- **Breadcrumbs**: This is implemented through {% block breadcrumbs %}, which includes a 'Breadcrumbs' template, providing navigation links to show the user's path to the current page.

- **Title**: {% block title %} contains an included Page Header template, displaying the page's title or other header content.

- **Main content and sidebar**: The template then defines a row (`div` with the `row` class), which is split into two main columns:

  - **Left column (main content)**: Occupies a larger portion (as indicated by the `col-sm-8 col-lg-8` class). It includes `{% block main %}`, where the `Page Copy` template is included, containing the primary content of the **Incident** page.

  - **Right column (sidebar)**: This is a smaller column (indicated by `col-sm-4 col-lg-4`) that houses `{% block aside %}`. This block is currently empty but can be used for additional content, such as side menus or widgets related to the incident details.

Overall, the template uses a two-column layout with a header section for breadcrumbs and a page title. It utilizes Liquid's block and includes functionality to modularize and inject different components, such as breadcrumbs, page headers, main content, and side content, providing a structured and flexible layout.

> **Important note**
>
> It's easier to copy an existing web template and make changes. Review the existing web templates provided by Power Pages.
>
> The author likes using the **Layout 2 Column wide** left web template as a starting point. So Sarah starts by copying that code and pasting it into the new incident web template.

Editing an existing web template in Power Pages code editor

Though developers need to create web templates in the Power Pages management tool, this is not a good place to work on them. It is more convenient to open the Visual Studio code editor from the Power Pages studio and edit web templates in the code editor, as shown earlier in *Figure 8.1*.

Defining the layout

In the web template editor, Sarah finds options to customize the header, body, and footer of her website by editing web templates. The header typically includes the site logo, navigation menu, and other branding elements. The body is where the main content of the web page resides and can include many web templates, while the footer contains additional information and links.

In the web template development process, developers have the flexibility to design and structure their pages using various tools and controls provided in Power Pages. Here's how to configure the layout of a web template:

- **Customization**: Developers have the ability to customize the sizing, alignment, and overall styling of these elements. This customization is key to achieving the specific look and feel desired for the web page, which uses either the inline code in the web template or in a separate CSS file.

- **Including standard components**: In the given example of *Figure 8.2*, Sarah includes standard components such as the page header web template.

Adding dynamic content with Liquid

One of the key advantages of web templates is their ability to incorporate dynamic content using Liquid, a powerful templating language. Liquid allows you to retrieve and display data from your Dataverse tables, manipulate the output, and implement conditional logic within your web template.

To add Liquid code to your web template, simply enclose it within `{% ... %}` tags. Within these tags, you can access Liquid objects, apply filters to modify the data and use control flow statements to conditionally display content or execute specific actions.

For example, you can retrieve a specific field value from a Dataverse table and display it in your web template using Liquid syntax such as `{{ entity.fieldname }}`. Liquid filters are used to modify the output of objects. They're separated from the value to which they're being applied by `|`. You can leverage Liquid filters, such as `| date` or `| capitalize`, to format the data before outputting it, for example, `{{ entity.fieldname | capitalize}}`.

Example for formatting a date to the UK date format

This Liquid syntax formats the date stored in `datefield` to the UK date format (day/month/year): `{{ datefield | date: 'dd/mm/yyyy' }}`.

Saving and applying the web template

Once you have finished customizing your web template, click the **Save** button to preserve your changes. You can then apply the web template to specific pages within your Power Pages.

To apply the web template, navigate to the desired page in the **Power Pages Management** app and select the web template from the available options. This will associate the web template with the selected page, ensuring that the defined layout and content are applied consistently.

By following these steps, you can create a basic web template in Power Pages. Remember to save your progress frequently and test your web template on various pages to ensure its functionality and appearance align with your desired outcomes.

Next, Sarah will explore the powerful capabilities of Liquid, which allows for dynamic content and data processing within your web templates.

Introduction to Liquid

Liquid is a templating language that plays a role in making web templates dynamic and flexible in Power Pages. Liquid is widely adopted across various platforms and provides a user-friendly syntax for processing and outputting data:

- **Data manipulation and display**: Utilize Liquid in Power Pages to manipulate and display data retrieved from Dataverse tables. This capability is essential to create dynamic web pages that reflect the latest data.

- **Content flow control**: Liquid acts as a bridge between the web template and the underlying data. This integration empowers control over the flow of content, allowing for the tailoring of the user experience based on specific data contexts.

- **Objects in Liquid**: Use Liquid objects to represent data elements, such as fields, from Dataverse tables. These objects enable access to and a display of their values, which is crucial for presenting accurate and relevant data on web pages.

- **Tags for logic and control flow**: Employ Liquid tags to manage control flow and implement logic. These tags enable conditional statements to iterate over collections of data, enhancing the interactivity and responsiveness of web pages.

- **Using filters to modify outputs**: Utilize Liquid filters to alter the output of variables, objects, or literals. This feature allows for the formatting of data as needed, ensuring that the information presented is clear and user-friendly.

- **Flexibility in web template design**: Liquid provides the flexibility to create dynamic content, handle complex data processing, and implement conditional logic within web templates. This versatility is key to developing sophisticated and user-engaging web pages.

This opens up a world of possibilities for creating personalized and interactive experiences for your users.

Getting started with Liquid in web templates

To begin using Liquid in web templates, Sarah needs to understand the syntax and structure of the language. Liquid code is enclosed within `{% ... %}` tags, and you can use various elements to achieve different outcomes. One of the fundamental aspects of Liquid is accessing objects. Use dot notation, such as `{{ object.property }}`, to retrieve specific properties or fields from objects. This allows Sarah to display data dynamically within a web template.

Control flow is another important aspect of Liquid. Use `if` statements, `for` loops, and other control flow tags to conditionally render content or iterate over collections of data. This enables dynamic sections within web templates to be based on certain conditions or to display data in a structured manner.

Liquid also provides a range of filters that can be applied to variables, objects, or literals. Filters modify the output of these elements, allowing for date formatting, string capitalizing, or other transformations. Filters are denoted by the pipe character (`|`) followed by the filter name, such as `{{ variable | filter }}`.

By understanding the syntax and structure of Liquid, Sarah can start incorporating it into her web templates to create dynamic and personalized experiences for users. In the following sections, Sarah explores how to get started with Liquid in web templates and leverage its capabilities to manipulate and display data.

In this example, Sarah is adding some JavaScript for an event handler to run on the `anyinjuries` field. Sarah does this by adding the JavaScript code within the main block of the `Page Copy` as follows:

```
{% block main %}
 {% include 'Page Copy' %}
```

```
<script type="text/javascript">
$(document).ready(function() {
  $("#imc_anyinjuries").change(AnyInjury);
  $("#imc_anyinjuries").change();
});
function AnyInjury(){
    var typeVal = ($("#imc_anyinjuries_1").is(":checked"));
    if(typeVal){
        SetFieldAsRequired("imc_injurydescription", "Injury
Description");
        }
}
</script>
    {% endblock %}
```

In this implementation, the code triggers the `AnyInjury` function whenever the state of the `imc_anyinjuries` checkbox changes. This function assesses whether the checkbox is checked, and if so, it sets the `"imc_injurydescription"` field as required. Thus, Sarah's code dynamically adjusts form field requirements based on user interactions, enhancing the user experience.

Transitioning from JavaScript integration, Sarah now shifts her focus to further explore the functionality of Liquid within Power Pages. This is a crucial step in understanding how to utilize Liquid objects to represent and access data within web templates. Liquid objects are instrumental in retrieving and displaying information from Dataverse tables, allowing for dynamic and personalized content creation.

In Power Pages, Sarah discovers a range of commonly used Liquid objects, each serving unique purposes. She is poised to delve into these objects, understanding how to effectively utilize them in her web templates. This exploration is pivotal in enabling Sarah to create more engaging and interactive Power Pages experiences.

Using Liquid objects in Power Pages

Liquid objects in Power Pages serve as gateways to data, allowing Sarah to craft dynamic and responsive web templates. These objects facilitate the retrieval and display of information from Dataverse tables, which is crucial for tailoring content to specific user needs. Here, Sarah will explore various Liquid objects and learn to apply them effectively:

- **Entity objects**: Entity objects represent records from Dataverse tables. For instance, accessing a field within an `entity` object is straightforward, as Sarah can use syntax such as `{{ entity. field_name }}`. If she's dealing with an `incident` record, displaying its title would be as simple as using `{{ incident.title }}`.

- **Current user object**: The current user object provides information about the currently logged-in user. Sarah can access attributes, such as username and email, by using syntax such as `{{ current_user.name }}`, enhancing personalization on her pages.

- **Site objects**: Site objects represent various properties of the Power Pages site. Sarah can use them to display elements such as the site name, `{{ site.name }}`.

- **Page objects**: Page objects represent the current web page being rendered. They provide access to properties such as the page title, URL, and parameters. For example, `{{ page.title }}` will display the title of the current page.

- **Request objects**: Sarah uses these to retrieve information from the current HTTP request. They're particularly handy for fetching query parameters or form data, such as `{{ request.querystring.param_name }}`.

By utilizing these Liquid objects, Sarah is well-equipped to create dynamic and personalized user experiences on her Power Pages. She's ready to implement these objects in her web templates, using them as placeholders to render corresponding data dynamically.

Let's consider an example of a web template for the incident details page. Sarah can use the `incident entity` object to access and display various fields of the incident record, such as the `title`, `description`, and `assigned user`. By using Liquid syntax, Sarah can incorporate these objects into her web template as placeholders, such as `{{ incident.imc_id }}`, to dynamically render the corresponding information. In this example, Sarah learns how to use the incident's ID as a page title in the web template for the page header, as shown here:

```
<header>
  <h1>{{ incident.imc_id | default: "Untitled Incident" }}</h1>
</header>
```

In this example, assume that `incident` is an object representing the current incident record in the web template. The `imc_id` property represents the ID of the incident.

The `{{ ... }}` double curly braces denote a Liquid variable. Sarah will use `incident.imc_id` to retrieve the value of the `imc_id` property. The default filter is used to provide a fallback value, `"Untitled Incident"`, in case the `imc_id` property is not present or is empty.

Sarah can place this code snippet within the `<header>` section of the web template to display the incident's ID as the page title.

After mastering the use of Liquid objects to dynamically display data such as incident IDs, Sarah now turns her attention to enhancing site navigation with breadcrumbs. **Breadcrumbs** are a fundamental component in web design, offering users a clear pathway to navigate through the different pages of a site. In Power Pages, Liquid plays a crucial role in creating these navigational elements.

Breadcrumbs web template explained

Sarah explores how to implement a `breadcrumbs` web template in Power Pages. Here's an example she reviews:

```
<nav>
  <ul class="breadcrumbs">
    {% assign breadcrumbs = page.breadcrumbs %}
    {% for breadcrumb in breadcrumbs %}
      <li><a href="{{ breadcrumb.url }}">{{ breadcrumb.label }}</a></li>
    {% endfor %}
  </ul>
</nav>
```

Note that `page.breadcrumbs` is an array of breadcrumb objects representing the hierarchical navigation path. Each `breadcrumb` object has two properties: `url` (the URL of the page) and `label` (the label to display `breadcrumb`).

Sarah can place this code snippet within a web template file to render the breadcrumbs navigation. The `<nav>` element provides semantic meaning for navigation, and the `<ul>` and `<li>` elements create an unordered list structure for the breadcrumbs. The Liquid `for` loop iterates over each `breadcrumb` object in the `breadcrumbs` array and generates a list item, `<li>`, with an anchor link, `<a>`, that points to the corresponding URL and displays the breadcrumb label.

To customize the breadcrumbs web template, you can modify the HTML structure, apply custom CSS styles to achieve the desired visual appearance and adjust the logic to match your specific navigation structure. For example, you can add additional classes or attributes to the HTML elements for styling purposes or extend the `breadcrumb` object to include additional properties such as an icon or tooltip.

Having learned how to use Liquid objects, Sarah will learn how Liquid tags work and their use in implementing logic and control flow within a web template in the next section.

Using Liquid tags in Power Pages

Liquid tags provide a mechanism for implementing logic and control flow within your web templates in Power Pages. They allow you to perform conditional checks, iterate over collections, and execute specific actions based on certain conditions. Let's explore some commonly used Liquid tags and how they can be utilized:

- If tag: The if tag allows you to perform conditional checks and executes specific code based on the evaluation of the condition. It follows the syntax `{% if condition %}...{% endif %}`. For example, you can use the if tag to display different content based on whether a certain field has a value or not:

```
{% if incident.title %}
    <h2>{{ incident.title }}</h2>
{% else %}
    <h2>No title available</h2>
{% endif %}
```

- for tag: The for tag enables an iteration over a collection, such as an array or a set of records, and performs actions for each item in the collection. It follows the syntax `{% for item in collection %}...{% endfor %}`. For example, use the for tag to display a list of related records:

```
<ul>
    {% for related_record in incident.related_records %}
        <li>{{ related_record.name }}</li>
    {% endfor %}
</ul>
```

- capture tag: The capture tag allows the storage of the output of a block of code into a variable for later use. It follows the syntax `{% capture variable_name %}...{% endcapture %}`. For example, use the capture tag to store the value of a field in a variable and use it multiple times within the template:

```
{% capture description %}{{ incident.description }}{% endcapture %}
<p>{{ description }}</p>
<p>{{ description | truncate: 100 }}</p>
```

- case/when tag: The case/when tag provides a way to handle multiple conditional cases within a single block of code. It follows the syntax `{% case variable %}...{% when value %}...{% endcase %}`. For example, you can use the case/when tag to display different messages based on the value of a field:

```
{% case incident.status %}
    {% when 'Open' %}
```

```
        <p>This incident is currently open.</p>
    {% when 'Closed' %}
        <p>This incident has been closed.</p>
    {% else %}
        <p>The status of this incident is unknown.</p>
{% endcase %}
```

These are just a few examples of the powerful capabilities of Liquid tags in Power Pages. By using these tags effectively, you can implement dynamic and responsive behavior within your web templates, enabling you to create customized experiences and handle various scenarios based on conditions and iterations.

> **Important note**
>
> Further reading on Liquid tags can be found here: `https://learn.microsoft.com/en-us/power-pages/configure/liquid/liquid-tags`.

Now, by having an understanding of how to use Liquid objects and tags, Sarah moves on to implement Liquid filters to transform her data.

Using Liquid filters in Power Pages

Liquid filters in Power Pages provide a way to modify and transform data within a web template. Filters allow Sarah to apply various operations on variables, objects, and strings to format and manipulate their output. Let's explore how filters can be used in Power Pages and provide an example using the `has_role` filter.

Filters are functions that take a value as input, perform a specific operation on it, and return the modified output. They are denoted by the pipe (|) symbol in the Liquid syntax. Filters can be used to **format dates**, **transform strings**, **manipulate arrays**, and much more. They help to achieve dynamic and customized content rendering within web templates.

Syntax for using filters

Understanding how to effectively use Liquid filters in Power Pages can significantly enhance the functionality and user experience of your site. Filters allow for the sophisticated manipulation of data and can be applied to various scenarios. In the upcoming example, Sarah learns how the `has_role` filter can be particularly beneficial in customizing user experiences based on their roles.

The syntax for using filters is `{{ value | filter_name: parameter }}`, where the value is the input to which the filter will be applied, `filter_name` is the name of the filter, and `parameter` is an optional parameter specific to the filter.

This example will demonstrate how to check user roles and dynamically display content, which is a common requirement in web applications. It is used when you want to set the visibility or required state of a field or other behavior on the page that depends on which web role or security level the user has. For example, an approval field may only be editable for admins or managers and read-only for everyone else.

Example – Using the has_role filter

The has_role filter is commonly used to check whether a user has a specific role assigned to them. It takes the roles assigned to the current user as input and returns a Boolean value, indicating whether the user has the specified role. It's an example of how Liquid can provide dynamic content based on user context: checking whether a user has a web role and then applying some logic if this is true. This pattern is widely used to manage form behavior, depending on whether a user has a security web role, to identify the type of user:

```
{% if user.roles | has_role: 'Admin' %}
    <p>Welcome, Admin!</p>
{% else %}
    <p>Welcome, Guest!</p>
{% endif %}
```

In this example, the has_role filter is used within an if statement to conditionally display different content based on whether the current user has the 'Admin' role. If the user has the 'Admin' role, the message Welcome, Admin! will be displayed. Otherwise, the message Welcome, Guest! will be displayed.

Commonly used filters

Power Pages provides a range of built-in filters that can be used in web templates. Some commonly used filters include the following:

- date: Formats a date string based on a specified format
- capitalize: Converts the first character of a string to uppercase
- truncate: Truncates a string to a specified length
- split: Splits a string into an array based on a specified separator
- join: Joins the elements of an array into a string using a specified separator

These filters, along with many others, can be used to manipulate and transform data within web templates.

> **Important note**
>
> Further reading for Liquid filters: `https://learn.microsoft.com/en-us/power-pages/configure/liquid/liquid-filters`.

By leveraging filters in Power Pages, you can customize and enhance the output of your web templates to meet specific formatting and data manipulation requirements. Experiment with different filters and their parameters to achieve the desired outcomes for your Power Pages.

Having learned the basics of Liquid and its syntax, we will introduce some advanced concepts in the next section.

Advanced Liquid concepts for Power Pages

In addition to the basic features of Liquid, Power Pages supports advanced concepts that allow for more dynamic and sophisticated web templates. These advanced concepts enable you to interact with data sources, iterate over collections, and perform complex operations. Let's explore some of these advanced Liquid concepts and provide code snippets to illustrate their usage.

Interacting with Dataverse

Power Pages allows you to interact with Dataverse, Microsoft's cloud-based storage system for PowerApps. This integration opens up possibilities for retrieving and manipulating data from Dataverse tables within your web templates. Here's an example of how you can fetch and display a list of data from a Dataverse table using Liquid:

```
{% assign incidents = entities.imc_incident %}
<ul>
  {% for incident in incidents %}
    <li>{{ incident.title }}</li>
  {% endfor %}
</ul>
```

In this example, the `entities.imc_incident` variable represents the Dataverse table "`imc_incident`". By iterating over each incident in the `incidents` variable, you can access and display specific properties of each incident, such as the title.

Iterating over collections

Liquid provides a powerful mechanism for iterating over collections, such as arrays or objects with multiple values. This allows you to display data dynamically based on the contents of a collection. Here's an example of how you can iterate over an array and display its elements using Liquid to count the number of incidents related to accidents:

```
{% assign incidents = incidents_table.records | where: "imc_
anyinjuries", true %}
{% assign incidents_count = incidents.size %}
<h2>Accident Incidents</h2>
<p>Total number of Incidents with Injuries: {{ incidents_count }}</p>
<ul>
  {% for incident in incidents %}
    <li>{{ incident.Name }}</li>
  {% endfor %}
</ul>
```

In this example, assume there is a table named `incidents_table` in Dataverse that contains incident records. Use the `where` filter to filter the records based on the `anyinjuries` field, retrieving only the incidents related to injuries. The filtered incidents are then assigned to the `incidents` variable.

Use the `size` filter to calculate the total number of accidents by getting the count of the incidents array and assigning it to the `incidents_count` variable.

The example then displays the total number of accidents and lists each incident's name using a `for` loop and the `li` HTML element.

This example demonstrates how to use Liquid to interact with Dataverse data, filter records based on specific criteria, and perform operations on the filtered collection. Developers can customize the logic and output to fit the specific requirements and data structure in Dataverse.

Performing conditional logic

Liquid provides conditional statements that allow the execution of different blocks of code based on specified conditions. This enables dynamic behavior within a web template. Here's an example of how to use conditional logic to display content based on a condition using Liquid:

```
{% if user.is_authenticated %}
  <p>Welcome, {{ user.display_name }}!</p>
{% else %}
  <p>Please sign in to access the content.</p>
{% endif %}
```

In this example, the `if` statement checks whether the user is authenticated. If the user is authenticated, a personalized welcome message is displayed with the user's display name. Otherwise, a message prompting the user to sign in is shown.

These advanced Liquid concepts provide flexibility to create more complex and dynamic web templates within Power Pages. By leveraging features such as interacting with Dataverse, iterating over collections, and performing conditional logic, developers such as Sarah can create highly customized and interactive experiences for their portal users. Experiment with these concepts and explore the extensive capabilities of Liquid to unlock the full potential of your web templates.

Now that Sarah has learned how to program in Liquid, she will examine ways to debug her code in the next section.

Debugging web templates and Liquid

Debugging is an essential part of the development process, and it helps identify and resolve issues in web templates and Liquid code. In this section, Sarah will explore some techniques and tools that can aid in debugging and troubleshooting.

Syntax errors

Ensure that Liquid code follows the correct syntax and structure. Mistakes in syntax can lead to errors when rendering the web template.

Check for missing or mismatched opening and closing tags, misplaced brackets or parentheses, and incorrect variable or filter usage, as shown in the following code block. In the provided code block, Sarah is learning the importance of correct syntax in Liquid, which is crucial for the error-free rendering of web templates. The example demonstrates a common error – a missing closing tag:

```
{% if condition %}
    <!-- Correct syntax -->
{% else %}
    <!-- Missing closing tag -->
{% endif %}
```

The `{% if condition %}` tag opens a conditional statement, but the corresponding closing tag, `{% endif %}`, is missing in the `'else'` section. This mistake can lead to issues in template processing, underscoring the need for a careful review of Liquid code to ensure all tags are properly opened and closed, maintaining the integrity and functionality of the web templates in Power Pages.

Logging output

Use the `capture` tag to store output into a variable and then output the variable's value for debugging purposes.

Print variable values or specific values within your web template to understand the flow of data, as shown here:

```
{% capture debug_output %}
    {% if condition %}
        {{ variable }}
    {% endif %}
{% endcapture %}
<!-- Output the captured debug output -->
{{ debug_output }}
```

Liquid tags and filters

When verifying the compatibility and behavior of tags and filters used in Power Pages, it's essential to confirm that they align with the intended functionality of the web template. The effectiveness of the Liquid tags and filters directly impacts the execution of the desired logic and data manipulation within the web template. Ensuring that these elements are suitable for Power Pages and behave as expected is crucial for the smooth operation and accuracy of the template's output.

You can check for the correct usage of parameters and the proper nesting of tags, as shown here:

```
{% if condition %}
    <!-- Correct tag usage -->
{% endif %}
{{ variable | filter }} <!-- Correct filter usage -->
```

Liquid objects and variable scope

You must confirm that you're referencing the correct objects and variables in your web template.

Further, you would need to be aware of the scope of variables and ensure they are accessible where needed, as shown in the following example. The {{ my_variable }} variable is defined within a certain scope due to the if condition. This demonstrates the need to consider where variables are defined to prevent issues with their accessibility in different parts of the web template:

```
{% assign my_variable = "Hello" %}
{% if condition %}
    {{ my_variable }} <!-- Accessible within the scope -->
{% endif %}
{{ my_variable }} <!-- May not be accessible outside the scope -->
```

Logging and error handling

Utilize the portal's logging and error-handling mechanisms to capture and review any errors or warnings that occur during the rendering of your web template.

Monitor the portal logs for any relevant information that can aid in debugging.

By employing these debugging techniques and being mindful of potential issues, you can effectively troubleshoot and resolve problems in your web templates and Liquid code, ensuring smooth rendering and functionality.

Remember, debugging is an iterative process, and it may require testing and refining your web templates and Liquid code until you achieve the desired results. Having introduced Liquid and looked at some code examples, Sarah will now have a quick look at best practices.

Best practices for using web templates and Liquid in Power Pages

When working with web templates and Liquid in Power Pages, the following best practices can greatly enhance the efficiency, maintainability, and performance of your portal. Here are some recommended practices to consider.

Consistent naming and structure

Use clear and descriptive names for your web templates, ensuring they reflect their purpose or functionality.

Maintain a consistent structure across your web templates, including common elements such as header, body, and footer. This promotes a cohesive user experience and simplifies maintenance.

Modular approach

Break down complex web templates into smaller, reusable components. This allows for easier management and updates and promotes code reusability.

Utilize "include" statements to include common code snippets or partial templates within your web templates. This reduces redundancy and enhances maintainability.

Commenting and documentation

Comment your Liquid code to provide clarity and context for future developers and yourself.

Document the purpose and functionality of your web templates, including any specific considerations or dependencies.

Performance optimization

Minimize the use of heavy computations or complex logic within your web templates. Excessive processing can impact the rendering speed and user experience.

Use Liquid filters judiciously to modify or format data only when necessary.

Leverage the caching mechanisms provided by the portal to improve performance for frequently accessed web templates.

Testing and validation

Regularly test your web templates across different browsers and devices to ensure consistent rendering and functionality.

Validate your Liquid code to catch any errors or potential issues early on.

Version control

Use a version control system to track changes and maintain a history of your web templates. This enables easy rollback and collaboration with other developers.

Security considerations

Be cautious when including user-generated content in your web templates to prevent security vulnerabilities.

Implement proper authentication and authorization mechanisms to protect sensitive data accessed through web templates.

Keep up with updates and best practices

Stay informed about the latest updates, features, and best practices related to web templates and Liquid in Power Pages. This ensures you are leveraging the most current and efficient techniques.

By following these best practices, you can optimize your development workflow, enhance the maintainability of your web templates, and deliver a high-quality and consistent user experience in your Power Pages.

Summary

We will now review the key points covered in this chapter:

1. Sarah learned the fundamentals of web templates in Power Pages, understanding their role in creating consistent and efficient web pages.

2. She explored Liquid, a templating language, and its application in making web templates dynamic and interactive.

3. Sarah delved into creating web templates, focusing on customization and adding dynamic content using Liquid.

4. Advanced Liquid concepts were explored, including interactions with Dataverse and complex data processing within web templates.

5. Debugging techniques for web templates and Liquid were covered, enhancing Sarah's problem-solving skills.

6. Best practices for using web templates and Liquid were discussed, focusing on efficient design and maintenance.

Sarah journeyed through the intricacies of web templates and Liquid in Power Pages. She grasped the foundational elements of web templates, recognizing their importance in creating standardized and cohesive web pages. With Liquid, Sarah tapped into the power of dynamic content rendering, learning to manipulate and display data from Dataverse tables and implement conditional logic in her templates.

Her exploration included hands-on experience in crafting web templates from scratch, where she personalized layouts and infused dynamic elements using Liquid. The advanced concepts of Liquid opened up new avenues for interacting with data sources, iterating over collections, and executing complex operations, enhancing her web pages' functionality and responsiveness.

Sarah's development skills were further sharpened by diving into debugging techniques, where she learned to identify and rectify issues in her templates and code. The chapter rounded off with a focus on best practices, guiding Sarah in adopting a structured, performance-oriented approach to web templates and Liquid use.

Equipped with this comprehensive understanding, Sarah is now adept at creating dynamic, efficient, and user-friendly web pages in Power Pages, ready to embrace the challenges of process automation in the next chapter.

9

Workflow Automation

In today's rapidly evolving and competitive business environment, organizations continually strive to streamline operations and enhance efficiency. Automation and well-defined processes are crucial for achieving these goals by eliminating manual tasks, reducing errors, and optimizing resource utilization. In this chapter, Sarah will explore the realm of processes and automation within the Power Pages framework.

In Power Pages, Sarah discovers two powerful tools for automating business processes: Dataverse workflows and Power Automate. Dataverse Workflow provides a built-in automation capability within Power Pages, enabling her to define and manage complex processes that span multiple steps and participants.

Sarah begins by examining the concept of workflow in Power Pages. She learns how workflows automate business processes within the website and understand the key components and associated terminology.

One example Sarah explores is the Power Pages form workflow button, available on Power Pages forms and lists. Sarah follows a step-by-step implementation of this workflow button, highlighting its automation and integration features.

In this chapter, Sarah will cover the following topics:

- Creating a notification with Dataverse Workflow
- Creating a Power Pages workflow button

By the end of this chapter, Sarah will be able to understand Dataverse Workflow in Power Pages.

Power Pages workflow

In Power Pages, a workflow refers to the automated sequence of steps that define and govern a business process within the website. It allows organizations to streamline their operations, improve efficiency, and ensure consistent and reliable execution of tasks.

Workflows enable Sarah to automate repetitive manual tasks, enforce business rules, and orchestrate the flow of information and actions.

Key components of a workflow

Here are the key components of a workflow:

- **Triggers**: Triggers define the events or conditions that initiate a workflow. They can be based on various actions within the website, such as creating or updating records or specific user interactions. Triggers act as the starting point for the workflow, determining when it should be activated.

- **Steps**: Steps represent individual actions or tasks within a workflow. Each step performs a specific operation, such as creating a record, sending an email, or updating a field. Workflows can consist of multiple steps that are executed sequentially based on the defined logic.

- **Conditions**: Conditions enable branching and decision-making within a workflow. Conditions allow developers to evaluate specific criteria and determine the next steps based on the outcome. Conditions can be used to create different paths or routes within the workflow, enabling t complex business processes with multiple possible outcomes to be automated.

- **Actions**: Actions are the operations that are performed by steps within a workflow. They can include creating, updating, or deleting records, sending notifications, assigning tasks to users, or invoking external services. Actions are the building blocks of workflow steps and define the operations to be executed.

Overall, workflows and automation in Power Pages empower organizations to automate a wide range of processes, ensure consistency, reduce manual errors, and enhance productivity.

The benefits of workflows in Power Pages

Let's look at some of the benefits of workflows:

- **Efficiency**: A workflow automates repetitive and time-consuming tasks, freeing up resources to focus on more valuable activities. It streamlines processes, reduces manual effort, and increases operational efficiency.

- **Consistency**: With workflows, Sarah can enforce standardized processes and ensure that tasks are executed consistently. This leads to improved quality and customer satisfaction.

- **Compliance**: Workflows allow organizations to enforce business rules, compliance regulations, and approval processes. They provide an auditable trail of actions and ensure adherence to predefined procedures.

- **Scalability**: As organizations grow, workflows enable them to handle increased volumes of tasks and maintain efficiency. They provide a structured approach to process execution, allowing for easy scaling and adaptability.

- **Collaboration**: Workflows facilitate collaboration and coordination among team members. They allow tasks to be assigned, progress to be tracked, and seamless handoffs between individuals or departments.

Power Pages provides a user-friendly interface for configuring and managing workflows. Sarah can define triggers, steps, conditions, and actions using an intuitive visual designer, eliminating the need for complex coding or technical expertise.

> **Further reading**
>
> You can learn more about classic Dataverse background workflows here: `https://learn.microsoft.com/en-us/power-automate/workflow-processes`.

In the next section, Sarah will learn how to use the Dataverse workflow provided by Microsoft with a Power Pages website installation.

Although Sarah planned to use Power Automate flows for most notifications (as explained in the next chapter), for this requirement, she decided to use the Dataverse Workflow email feature.

Creating a notification with Dataverse Workflow

Sarah needs a simple notification to inform management that an incident has been created and send it directly to both Brenda and Rob, who want to know immediately if there is an incident on any site. Sarah decides to trigger the workflow when an incident is created. Although she's considered using the status reason of an incident, she determines there's a risk that the incident reporter might not update the status promptly. Management expressed that they want immediate notifications.

As usual, Sarah begins with an Agile user story to define the use case.

Agile user story – creating a notification with Dataverse Workflow

As a system administrator, I want to create a workflow using Dataverse to notify management immediately when an incident is created so that they can respond promptly to any urgent issues.

Description

Sarah, a system administrator, needs to set up an automated notification system within Power Pages using Dataverse Workflow. This system will send an email to the management team (Brenda and Rob) whenever a new incident is reported. The notification must be triggered by the creation of an incident to ensure immediate awareness. The workflow will specifically notify management about accidents and not employee write-ups to prioritize urgent incidents.

Acceptance criteria

Here are the acceptance criteria:

1. Workflow creation:

    - The workflow should be created within the Buildapp solution
    - The workflow should be named `Notify Management for a new Incident`
    - The workflow should not run in the background (real-time instant workflow)

2. Trigger setup:

    - The workflow should be triggered when a new incident record is created

3. Scope and conditions:

    - The workflow should have an organizational scope to ensure it runs for all users
    - A condition should be set to check if the incident type is **Accident**

4. Email notification:

    - An email should be sent from Brenda's account (as she has the necessary permissions)
    - The email should be sent to both Brenda and Rob
    - The email's content should include relevant details about the incident, including fields from the incident record (for example, the originator's name and telephone number)

5. Testing:

    - The workflow should be tested by creating a test incident
    - The test email should be confirmed to be received by Brenda and Rob

Tasks

To implement this story, Sarah must do the following:

1. Access Power Pages at `https://make.powerpages.microsoft.com/`.
2. Navigate to the **Solutions** tab and select the Buildapp solution.
3. Create a new workflow under the **Automation** section.
4. Name the workflow `Notify Management for a new Incident` and set it to run instantly.
5. Set the workflow to trigger the creation of an `incident` record.
6. Set the workflow's scope to **Organization**.

7. Add a condition to check if the incident type is **Accident**.

8. Create an email notification step with the following details:

   From: Brenda

   To: Brenda and Rob

   Email content: Include incident details using the form assistant

9. Set the **Regarding** field to the **Incident** table.

10. Save and activate the workflow.

11. Test the workflow by creating a test incident and verify that the email is sent correctly.

Definition of done

Here are the definitions of done details:

- The workflow has been created, configured, and activated in the Buildapp solution

- An email is sent to Brenda and Rob when a new incident of the **Accident** type is created

- The workflow has been tested and verified that it functions as intended

Notes

The following are some items to consider:

- Ensure all necessary permissions are set for sending emails within Dataverse

- Verify the accuracy of incident details included in the email notification

This user story encapsulates the process Sarah needs to follow to create and test the Dataverse workflow for incident notifications, ensuring management is promptly informed about critical incidents.

Workflow implementation

In this section, Sarah will create the workflow. To organize her work and make the workflow easy to deploy, Sarah will create the workflow within her **Buildapp** solution. Sarah creates the workflow by following these steps:

1. First, Sarah opens https://make.powerpages.microsoft.com/.

2. Then, she selects the **Solutions** tab.

3. Next, she selects the **Buildapp** solution. This solution contains most of Sarah's Dataverse development work.

4. At this point, Sarah selects a new workflow by going to + **New | Automation | Process | Workflow**, as shown in *Figure 9.1*:

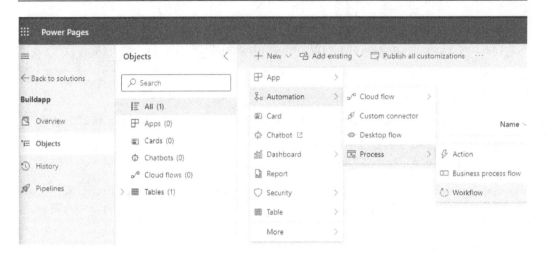

Figure 9.1 – Selecting a new workflow

5. Upon opening a new workflow, Sarah chooses whether to use a background workflow or not (which then creates a real-time instant workflow) and gives it a meaningful descriptive name. Sarah enters `Notify Management for a new Incident` as the workflow's name and disables **Run workflow in the background** so that it is an instant workflow, as shown in *Figure 9.2*:

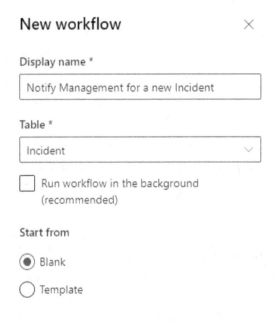

Figure 9.2 – New workflow settings

6. As shown in *Figure 9.2*, Sarah must select **Incident** under **Table** so that she can configure the incident trigger and fields she must fill in as part of the workflow automation process. Sarah presses the **Create** button, which then creates the workflow and opens the workflow designer.

Sarah changes the scope of the workflow so that it runs under the organization and not under the user, which could restrict the organization when the workflow gets triggered. Sarah checks that the trigger is set to **Record is created** so that it will run when the incident is submitted from the new incident web page; this is when the incident record is first created.

Next, Sarah must add some steps to the workflow. She remembers that she learned it is always best to start a workflow with a condition. When discussing how this should work with Brenda, she remembers the following: Brenda told her they only want to be notified of accident incidents but not employee write-ups as those aren't urgent and can be dealt with as part of appraisals and regular meetings with the site managers.

Sarah had set up two types of incidents: **incident** and **employee write ups**. Accidents were planned to be of the incident form type and Brenda had explained that accident incidents were important because emergency services may be involved and there is a statutory requirement to inform authorities of onsite accidents. Also, most of the project agreements with clients involved commitments to inform clients of onsite accidents.

Therefore, Sarah had to check that the incident was of the **Accident** type, as shown in *Figure 9.3*. Sarah continued to complete the workflow by doing the following:

1. Sarah, Sarah adds a conditioning step where **Form type** equals **Incident**.
2. Under **Add Step** within the condition, Sarah adds **Send email** as a step with **Create New Message** as its option, as shown in *Figure 9.3*:

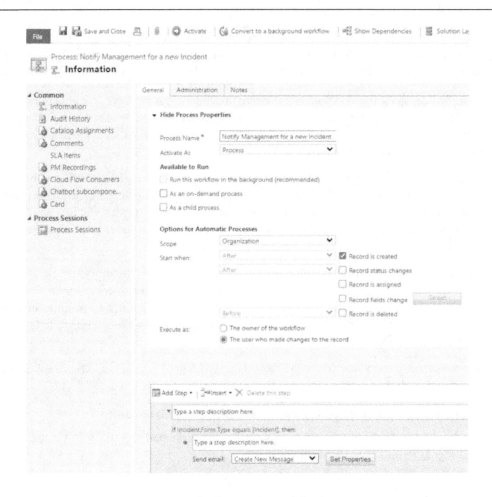

Figure 9.3 – Configuring the workflow's steps

3. Then, Sarah selects that step's **Set Properties**, which opens the **Send Email** form.

4. In the **From** field, Sarah selects **Brenda Smith** as Brenda has an Outlook license and is allowed to send and receive emails within Dataverse.

5. In the **To** field, Sarah selects both **Brenda Smith** and **Robert Smith** so that they both get the email.

6. Now, Sarah enters some text. Using the form assistant, she can add fields from the incident record, as shown in *Figure 9.4*. The fields are highlighted by the workflow designer in yellow and all fields are available from both the incident record and related lookups. This means that Sarah could add the name and telephone number of the originator, plus other information.

7. The **Regarding** field shows where the email will be referenced and it defaults to the referenced table, which is the **Incident** table, so that the email will be stored under the incident that's being created.

8. Sarah saves the email message and activates the workflow.

9. Sarah tests it by creating a new test incident and checks that the email was created. She lets both Brenda and Rob know to ignore the test incident email:

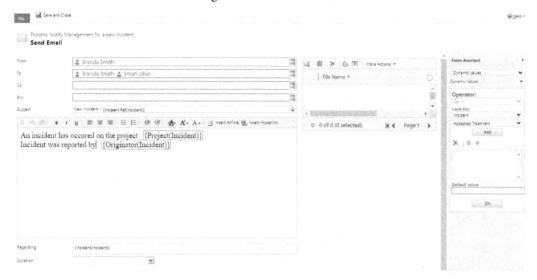

Figure 9.4 – Creating an email

The email was successfully tested and it was really quick to create and test. Sarah did consider adding other information to the incident record as she realized that at this point, she could add more information to the incident record using the same workflow if needed. For example, Sarah could add an update record step to set information on the incident record.

Brenda said that she was satisfied with a simple notification and that management could follow up. In the next section, we'll discuss a form button that would involve creating another workflow to set the incident's status.

Creating a Power Pages workflow button

Now, Brenda has said that she wants a button on the incidents page that will set the incident's status when it's completed and submitted by the incident originator. The originator, in most cases, will be the site manager or project manager. Brenda and Sarah agree to call the button **Approve Submission**. Sarah thinks about this and decides the easiest way is to use a workflow button, which she can place on the incident web page's basic form and also have the button filtered for its visibility. Sarah writes out an Agile user story for the use case.

Agile user story – creating a workflow button for incident status

As a site manager, I want to have a button on the incident page to approve the submission so that I can easily update the status of an incident to **Approved Submission**.

Description

Brenda wants a button on the incident form page to allow site managers and project managers to set the incident status to **Approved Submission** when an incident is completed and submitted. This will help in efficiently managing and tracking the status of incidents without manual updates.

Acceptance criteria

Here are the acceptance criteria for this user story:

1. Workflow creation:

    *   A new workflow named `Set incident to Approved Submission` must be created on the **Incident** table
    *   The workflow should be configured as an on-demand process
    *   The scope of the workflow should be set to **Organization**
    *   The workflow should change the incident's status to **Approved Submission**

2. Button configuration:

    *   A button labeled **Approve Submission** must be added to the incident form
    *   The button should only be visible for incidents of the **Incident** type and not for **Employee Write Up**

3. Button settings:

    *   The button should have a tooltip stating **This will set the incident to a status of Approved Submission**
    *   The button should be placed at the top right of the form

4. Filtering visibility:

    *   The button's visibility should be filtered using an advanced find tool to ensure it only appears for the specified incident type

5. Testing:

    *   The button functionality should be tested to confirm that it correctly sets the incident status to **Approved Submission**

Tasks

To implement this story, Sarah must do the following:

1. First, she must access the Buildapp solution in Power Pages.

2. Then, she must check and add a status reason of **Approved Submission** to the **Incident** table.

3. Next, she needs to create a new workflow named `Set incident to Approved Submission` on the **Incident** table.

4. After, she must set the workflow as an on-demand process with an organizational scope.

5. At this point, Sarah must add a step in the workflow to change the incident's status to **Approved Submission**.

6. Now, she must open the basic form of the **Incident edit** page.

7. Sarah needs to select the **Additional Settings** tab.

8. Under **Actions for Workflow**, Sarah must select the **Set incident to Approved Submission** workflow.

9. Sarah must configure the button settings as follows:

   - **Label**: `Approve Submission`

   - **Tooltip**: This will set the incident to a status of **Approved Submission**

   - **Placement**: Top right of the form

10. Sarah must also add a filter for the button so that it's only visible on the incident form and not on the employee write-up form.

11. Finally, Sarah must test the button by updating an incident and verifying that its status changes to **Approved Submission**.

Definition of done

Here are the definitions of done details:

- The **Set incident to Approved Submission** workflow is created and configured correctly

- The **Approve Submission** button is added to the incident form and functions as intended

- The button is only visible for incidents of the **Incident** type

- The button's tooltip and placement are configured as specified

- The workflow and button are tested and verified to work as expected

Notes

The following are some items to consider:

- Ensure all necessary permissions and configurations are in place for the workflow to function correctly
- Document the workflow and button setup for future reference and maintenance

This user story outlines the steps Sarah needs to follow to create a workflow button for updating the incident's status, ensuring efficient and accurate status management for incidents.

Workflow implementation

Sarah needs to create the workflow as an on-demand workflow and then add a button to the incident basic form. To do so, she does the following:

1. First, Sarah browses to the **Buildapp** solution in Power Pages.
2. Sarah checks the **Incident** table's status reason and adds a status reason of **Approved Submission**.
3. Then, Sarah creates a new workflow on the **Incident** table called `Set incident to Approved Submission`.
4. Next, Sarah sets the **As an on-demand process** workflow and sets the scope to **Organization**.
5. Sarah only needs to do one more thing here: she must set the change status to **Approved Submission**.
6. Next, Sarah opens the basic form of **Incident edit**.
7. She selects the **Additional Settings** tab, as shown in *Figure 9.5*.
8. Then, she selects **Actions** for **Workflow** and selects the **Set incident to Approved Submission** workflow.
9. Sarah makes the following configuration changes:

    I. She sets **Button label** to `Approved Submission`.
    II. She sets **Button Tooltip** to `This will set the incident to a status of Approved Submission`.

III.   She sets button placement and alignment to the top right of the form.

IV.   She also adds a filter as she only wants this button to be visible on **Incident** forms and not on **Employee Write Up** forms:

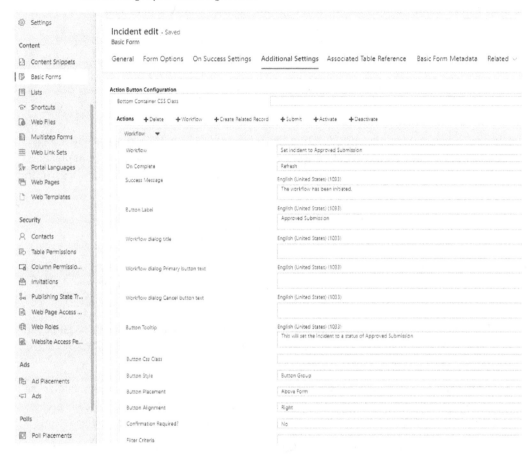

Figure 9.5 – Additional Settings – Action Button Configuration

Using the **Advanced Find** tool to help her, Sarah gets the following filter:

```
<fetch version="1.0" output-format="xml-platform" mapping="logical" >
  <entity name="imc_incident">
    <attribute name="imc_incidentid" />
    <filter type="and">
      <condition attribute="imc_formtype" operator="eq"
value="176230000" />
    </filter>
  </entity>
</fetch>
```

Now that Sarah has explored the customization capabilities of Dataverse Workflow in Power Pages, she is considering some best practices for processes and automation. By following these best practices, Sarah can ensure the effectiveness, reliability, and maintainability of automated processes within Power Pages.

Best practices for processes and automation

When working with processes and automation in Power Pages, it's important to adopt best practices to maximize the benefits and minimize potential challenges. Here are some recommended best practices to consider:

- **Clearly define objectives**: Before implementing any process or automation, clearly define the objectives and expected outcomes. Understand the purpose, scope, and desired results of the automation to ensure it aligns with the business goals.

- **Analyze and streamline existing processes**: Before automating a process, thoroughly analyze the existing manual process to identify areas of inefficiency or bottlenecks. Streamline the process as much as possible before introducing automation to optimize efficiency and effectiveness.

- **Start simple and iterate**: Begin with simpler automations and gradually build upon them. Start with small, manageable processes and workflows, and then iterate and enhance them based on feedback and evolving requirements. This iterative approach allows for continuous improvement and avoids overwhelming complexity.

- **Involve stakeholders**: Engage relevant stakeholders throughout the automation process. Collaborate with process owners, end users, and subject matter experts to ensure the automation meets their needs and addresses any specific requirements or challenges.

- **Design for scalability and flexibility**: Consider the future scalability and flexibility of automated processes. Anticipate potential growth and changes in the organization and design the workflows so that they accommodate evolving needs. This includes considering dynamic data sources, configurable parameters, and extensibility options.

- **Standardize naming conventions and documentation**: Establish consistent naming conventions for workflows, actions, and variables to enhance clarity and maintainability. Document the purpose, logic, and configurations of each automation component to facilitate understanding and troubleshooting.

- **Test rigorously**: Thoroughly test automated processes to ensure they function as intended. Conduct testing in various scenarios to identify and address any potential issues or unexpected behaviors. Perform regular regression testing when making changes or updates to workflows and pages.

- **Monitor and measure**: Implement monitoring and tracking mechanisms to gain insights into the performance and effectiveness of automated processes. Monitor key metrics, such as completion times, error rates, and user feedback, to identify areas for optimization and improvement.

- **Security and compliance**: Consider security and compliance requirements when designing and implementing automated processes. Ensure appropriate access controls, data protection measures, and compliance with relevant regulations and policies.

- **Regular maintenance and updates**: Automated processes require ongoing maintenance and updates. Regularly review and optimize the workflows while addressing any issues, incorporating user feedback, and accommodating evolving business needs.

By adhering to these best practices, Sarah can create robust and efficient processes and automation within Power Pages. These practices help ensure the success of automation initiatives, improve user experiences, and drive operational excellence within the organization.

Now, let's recap what Sarah has learned in this chapter.

Summary

In this chapter, Sarah delved into the world of processes and automation within the context of Power Pages. Sarah explored how workflows empower organizations to streamline operations, automate tasks, and improve efficiency.

Sarah learned about the key components of Dataverse Workflow in Power Pages, including triggers, steps, conditions, and actions.

Sarah also explored the importance of involving stakeholders, testing rigorously, monitoring performance, and ensuring security and compliance.

By the end of this chapter, Sarah gained an understanding of Dataverse Workflow in Power Pages. She is now equipped with the knowledge and tools to implement Dataverse Workflow to optimize business operations.

In the next chapter, Sarah will continue with process automation by learning how to integrate Power Automate flows with Power Pages.

10
Power Pages and Cloud Flows

Sarah's enthusiasm for Power Automate was fueled by her discovery of Power Pages connectors, which promised instantaneous data retrieval for use on a page. Engaging in extensive discussions with her client about implementing notification flows triggered by record changes, such as status updates, she delved into tutorial videos and learning materials. However, Sarah recognized that practical experience was essential to truly grasp the platform's capabilities. Exploring two avenues for process implementation on Power Pages, she envisioned using flows triggered by Dataverse record changes, such as status updates on incident records, to notify relevant individuals. Alternatively, she considered embedding flows directly into the Power Pages site, invoking them from JavaScript to execute processes and potentially return values for dynamic page rendering. Sarah's ideas aimed to leverage Power Automate effectively within the Power Pages environment, demonstrating her proactive approach to mastering the platform.

As Sarah delved deeper into the realm of Power Automate flows, she realized its pivotal role as a core skill, tightly intertwined with Power Pages. Understanding its significance, she recognized that Power Automate not only facilitates seamless integration with other systems but also empowers automation within the pages themselves.

Here's a summary of what Sarah knows about Power Automate thus far:

- Power Automate stands as a robust and extensible automation platform, intricately connected with Power Pages, offering advanced capabilities for process automation and seamless integration with various systems.

- Exploring Power Automate further reveals its potential to enhance workflow capabilities within Power Pages, enabling more sophisticated automation scenarios and providing extensive customization options tailored to specific business needs.

- Power Automate serves as a potent automation tool, seamlessly integrating with Power Pages to extend its functionality and enable advanced automation possibilities. With its flexibility, users can create intricate workflows, establish connections with external systems, and fine-tune automation processes to align with unique business requirements.

In this chapter, we'll cover the following topics:

- The key features of Power Automate in Power Pages

- Integration with Power Automate

- Use case – automating email notifications for timesheet approvals

- Cloud flow triggered by Dataverse

- Cloud flow integrated with Power Pages

The key features of Power Automate in Power Pages

Here are the key features of Power Automate:

- **Workflow designer**: Power Automate provides a visual workflow designer that allows you to create and manage process flows with ease. The designer offers a wide range of triggers, actions, and conditions that you can drag and drop to build your automation logic. It offers a user-friendly interface that empowers both technical and non-technical users to create complex workflows.

- **Connectors**: Power Automate offers a vast library of connectors that enable seamless integration with various external systems, services, and data sources. These connectors allow you to interact with applications such as SharePoint, Dynamics 365, Microsoft Teams, and many more. By leveraging connectors, you can automate interactions between Power Pages and other systems, enhancing data exchange and process efficiency.

- **Custom actions**: Power Automate enables the creation of custom actions that extend the capabilities of your workflows. These actions can be developed using code or low-code tools such as Power Automate Desktop. Custom actions allow you to perform specific tasks, invoke APIs, or execute complex business logic within your workflows.

- **Approval processes**: Power Automate provides built-in capabilities for creating approval processes within your workflows. You can configure multi-level approvals, define approval criteria, and customize notifications. This feature is particularly useful for scenarios such as document approvals, purchase requests, or expense claims.

- **Data transformations**: Power Automate allows you to manipulate and transform data within your workflows. You can use built-in functions and expressions to perform calculations, apply formatting, extract specific values, or merge data from different sources. This capability enables you to enrich and refine the data flowing through your automation processes.

- **Error handling and logging**: Power Automate offers comprehensive error handling and logging mechanisms. You can define error-handling steps, configure retries, and set up notifications for failed workflow instances. Additionally, you can log workflow execution details to monitor performance, troubleshoot issues, and gain insights into process optimization.

> **Further reading**
> Power Automate: `https://learn.microsoft.com/en-us/training/powerplatform/power-automate`

Power Automate in Power Pages provides seamless integration between your website and other systems or services. By leveraging Power Automate, you can create end-to-end automation scenarios that encompass data retrieval, manipulation, integration, and notification, all within the context of Power Pages.

The flexibility and extensibility of Power Pages, along with its integration with Power Automate, provide a robust platform for creating sophisticated workflows that streamline business operations and enhance productivity within your website environment.

Integration with Power Automate

Power Pages seamlessly integrates with Power Automate, Microsoft's low-code/no-code automation platform. This integration expands the capabilities of your workflows, allowing you to leverage the vast array of connectors, actions, and data transformation capabilities available in Power Automate. By integrating Power Automate with Power Pages, you can extend your workflow functionality, connect to external systems, and automate complex business processes that span beyond the website itself.

Power Automate expanded its capabilities with its visual designer, connectors, custom actions, and data transformations. Through customization, Sarah tailored workflows to match specific business requirements and enhance the functionality of Power Pages.

Sarah collaborated with her client to establish a comprehensive timesheet system within the Dataverse environment. Adam, a Dataverse developer in the accounts department, had laid the foundation by creating the necessary tables and utilizing the standard model-driven app. However, Sarah identified the need for additional support, particularly with Power Pages integration, to enhance the system's functionality. Leveraging her expertise, Sarah assisted the client by designing and implementing user-friendly pages for timesheet entry and submission. Now, the client has asked Sarah to integrate an email notification to inform the client that the timesheet is ready for approval.

Use case – automating email notifications for timesheet approvals

Title: Automating Email Notifications for Timesheet Approvals.

As a: Power Pages developer.

I want: To automate email notifications for timesheet approvals using Power Automate.

So that: Approvers are immediately informed when a timesheet is ready for approval, streamlining the approval process and enhancing communication efficiency.

Acceptance criteria

Let's look at this use case's criteria:

1. The system should trigger an email notification when a timesheet's status changes to "awaiting approval."
2. The email should contain a PDF attachment of the timesheet and a link to the timesheet page.
3. The flow should handle errors gracefully and log necessary details for troubleshooting.
4. The flow should be reusable for other notification requirements with minimal modifications.

Cloud flow triggered by Dataverse

Sarah learned how to trigger a cloud flow via a Dataverse record being modified, Here, Sarah thought a status change was an appropriate trigger for what was needed. Sarah engineered an automated system using cloud flows to dispatch email notifications to the client upon timesheets being ready for approval, thereby augmenting communication channels and expediting the approval process seamlessly. Sarah envisioned reusing this pattern to fulfill her client's broader notification requirements; they had a substantial list of notifications and it meant that Sarah could **Save As** on the flow and easily create new notification flows with the same pattern based on this cloud flow.

Among various connectors available for email dispatch in Power Automate Flow, Sarah opted for **Send an Email (V2)** due to its compatibility with the client's Outlook license and its simplicity in attaching files, aligning with their preferences and ease of use.

Designing a notification cloud flow

Sarah's next task was to develop a notification flow to inform the client when a timesheet was ready for approval. The objective of this notification was to alert the approver that a timesheet was awaiting approval, indicating that the worker had completed their entries and submitted them. The email notification was to include an attachment of the timesheet PDF, generated separately and stored in a file field on the timesheet record. The cloud flow is triggered by a Dataverse connector when the timesheet's status reason field changes, specifically when the status is updated from **open to awaiting approval**.

Implementation

Sarah initiated the development process by accessing her solution from Power Pages, her preferred method for managing objects within the environment. To create a new Power Automate flow, Sarah navigated to the **Cloud flows** tab in the solution and selected **New Automation** to begin the **Cloud flow** creation process, as illustrated in *Figure 10.1*:

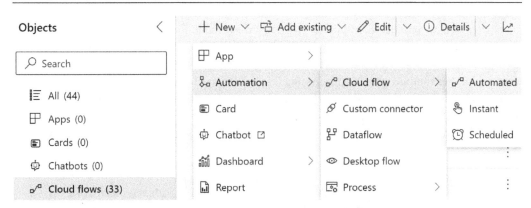

Figure 10.1 – Creating a new cloud flow

On selecting a new automated cloud flow, the connectors were displayed, Sarah chose Dataverse and the **When a row is added, modified, or deleted** option, and used `Notify - Client Timesheet requires Approval` as the name.

Sarah configured the first step to trigger on a modified timesheet when a status changes to awaiting approval, as shown in *Figure 10.2*:

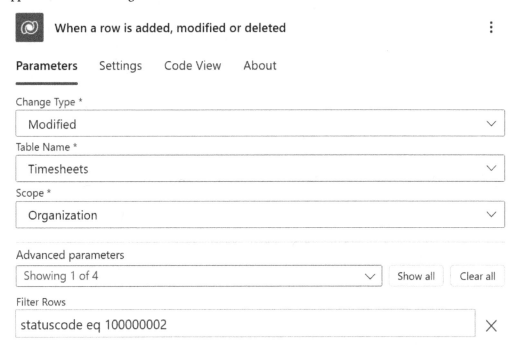

Figure 10.2 – A Power Automate cloud flow trigger

It's important to have a proper filter flow setting so that the cloud flow only triggers the status change to awaiting approval. To get the correct field names and values, Sarah had to look for the **status reason** field and the options for that field to get the correct value. Sarah did this by adding the **status reason** field to her solution and then examining it. Another way to get the exact value would be to use **advanced find** to find a timesheet where `status reason eq awaiting approval`, download the **fetchxml**, and then copy the values out of the advanced find. Sarah got in the habit of adding columns that she referred to in her solution so that it was easy for her to examine.

> **Further reading**
>
> You can learn more about advanced find here: `https://learn.microsoft.com/en-us/power-apps/user/advanced-find#create-edit-or-save-a-view-using-legacy-advanced-find`.

Sarah completed the flow steps, as shown in *Figure 10.3*. She also used Copilot to help with this by offering various prompts, such as Get row by ID and then sending an email with attachments, with varying degrees of success. In any case, Sarah knew what she wanted to do – at least Copilot gave her the send email step, which she used:

Figure 10.3 – Complete cloud flow steps

Sarah wanted to use fields from three related tables. The easiest way to do this, while also making it convenient to test the flow, was for her to add each related table as a **Get Row by ID** step while using values from the timesheet, as shown in *Figure 10.4*:

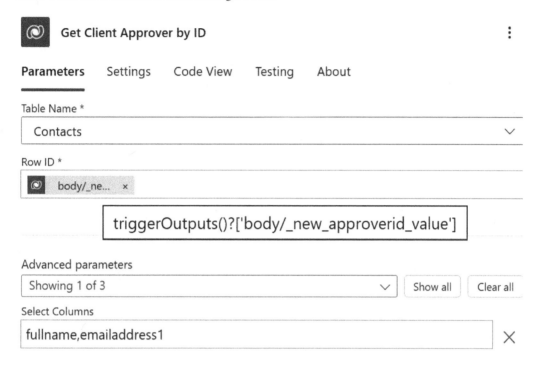

Figure 10.4 – The Get Row by ID step

Here, there's a **Contacts** table as Sarah is looking to get the name of the approver. **Row ID** is the approver lookup in the timesheet with a column name of new_approverid. Sarah only wants to retrieve the full name and email address of the approver, so she adds fullname and emailaddress1 to **Select Columns**.

This makes it easy for Sarah to create an email text with dynamic content from the timesheet, as shown in *Figure 10.5*:

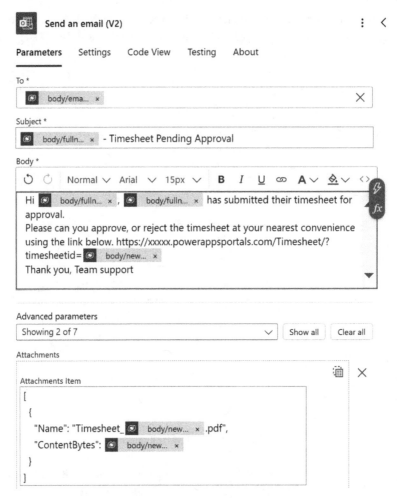

Figure 10.5 – Email content

By clicking in the fields for **To** and **Subject**, Sarah can select data from any of the steps. Here, Sarah selects the approver's email address and full name, respectively. Then, Sarah enters text and selects step data as needed, including a link to the website page to open the timesheet pages. Sarah also configures the PDF as an attachment. The PDF exists on the timesheet record as a file field named **PDF**. Here's the syntax for adding an image or file as an attachment:

```
[
  {
    "Name": "Timesheet_@{triggerOutputs()?['body/new_name']}.pdf",
    "ContentBytes": @{triggerOutputs()?['body/new_pdf']}
  }
]
```

Sarah tested this by setting herself up as an approver on a test timesheet and triggering the flow by changing its status to awaiting approval. Next, Sarah had a more complex Power Automate cloud flow that involved registering the flow within Power Pages as an instant flow.

Cloud flow integrated with Power Pages

Power Automate cloud flows can be integrated directly into Power Pages so that the flow can be triggered by a change on the page. At this point, its response results can be set on the page. Sarah thought of many ways she could use this pattern that previously had to be done by Dataverse plugins or complex JavaScript, which was beyond Sarah's programming capabilities.

Sarah had been asked to use a public service that will retrieve a company's registered address from the company registration number from the client account page and populate the client address. This was needed because her client had been carrying out credit checks but if the company's registered address was not accurate – it sometimes gave inaccurate results and cost her client wasted time and effort.

Design

The design aims to leverage Power Automate cloud flows within Power Pages to streamline processes previously requiring Dataverse plugins or complex JavaScript. This integration enables flows to be triggered from page changes and response results to be displayed on the page. Sarah, facing the challenge of retrieving accurate company addresses for credit checks, saw this integration as a solution. By integrating a public service API into the client account page, Sarah aimed to automatically populate the client's address based on the company registration number. This automation would save time and effort, reducing inaccuracies in credit checks.

To implement this solution, Sarah followed a structured approach. First, she created a cloud flow named `Get Company Registered Address` within Power Pages Studio. Configuring this flow involved defining a Power Pages trigger step for the company number input and subsequent steps to invoke the public service API, parse the JSON response, and output the address fields. Once configured, Sarah integrated the flow into the web template's JavaScript. She leveraged existing Microsoft examples, modifying the code to fit her requirements. The JavaScript code retrieves the company number input, constructs the cloud flow URL, and executes a `POST` request to trigger the flow. Upon successful execution, the response data is parsed, and address fields on the page are populated accordingly.

This design aligns with Sarah's goal of automating address retrieval within Power Pages, enhancing efficiency and accuracy in credit check processes. By seamlessly integrating Power Automate cloud flows, Sarah can achieve automation without extensive programming knowledge, empowering her to meet her client's needs effectively.

Implementation

Sarah had been given GET as the URL from the public service API.

The URL had a parameter of {companyNumber}, which was a field on the client account form, as shown in *Figure 10.6*:

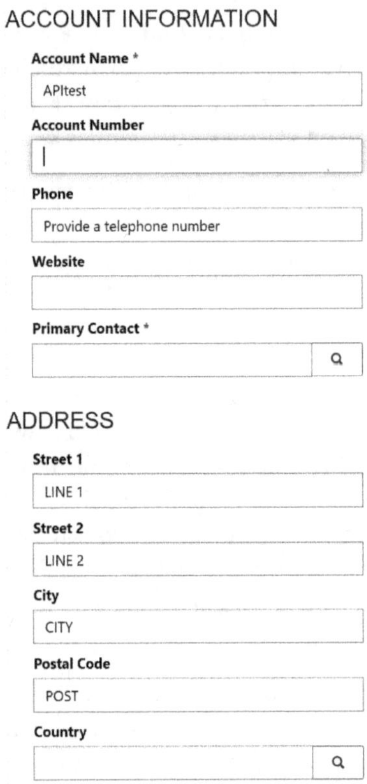

ACCOUNT INFORMATION

Account Name *

APItest

Account Number

|

Phone

Provide a telephone number

Website

Primary Contact *

Q

ADDRESS

Street 1

LINE 1

Street 2

LINE 2

City

CITY

Postal Code

POST

Country

Q

Figure 10.6 – Form integrated with cloud flow

The **Account Number** field will contain the company number required by the public service API to retrieve the address.

The cloud flow, together with JavaScript within the form, needs to retrieve the company address and fill the fields of the **ADDRESS** section, from **Street 1** to **Postal Code**. To complete this, Sarah needed to perform the following steps:

1. Create the cloud flow to GET the HTTP request returning the address.
2. Implement the flow into the JavaScript of the form, specifically its web template.
3. Extract the results from the flow's JSON response and set them to the **ADDRESS** fields.

The first thing Sarah needs to do is create the cloud flow. Sarah opens Power Pages Studio and selects **Set up**, then **Cloud flows**, as shown in *Figure 10.7*:

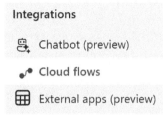

Figure 10.7 – The Cloud flows tab in Power Pages Studio

This tab opens the **Cloud flows** configuration page, where users can add new cloud flows or add existing flows to be configured with this Power Pages website:

Figure 10.8 – Creating a new cloud flow

Sarah will create a new cloud flow and select the **Power Pages** connector to do so, as shown in *Figure 10.9*:

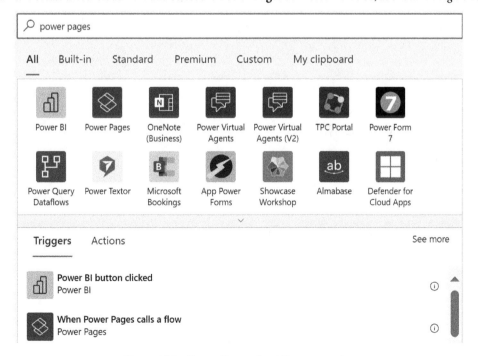

Figure 10.9 – Power Pages cloud flow connector

Sarah called the flow `Get Company registered address` and configured it as shown in *Figure 10.10*:

1. On the **Power Pages** trigger step, she added text input for the company number.

2. Then, she selected the **HTTP GET** step and entered the `GET API URL`, replacing the placeholder for the company number with the input variable of the company number.

3. Then, Sarah added a **Parse JSON** step to parse and extract the data into usable fields. The schema she entered uses sample data contained in the API documentation:

Figure 10.10 – Cloud flow steps for a GET HTTP request

Finally, Sarah added a **PowerApp Response** step, which, as shown in *Figure 10.11*, outputs the text fields as variables that can be referenced in the JavaScript code that calls the flow:

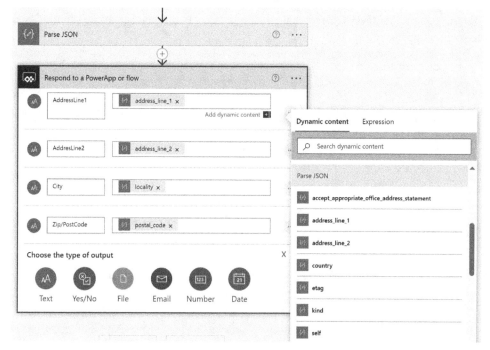

Figure 10.11 – Cloud flow steps to output address fields

With that, Sarah can consume the flow in the JavaScript of the web template.

Web template JavaScript

Sarah decides to look at the Microsoft example in Microsoft Learning, copy the JavaScript example, and modify that code to suit her requirements. The web template will use the safe Ajax that she installed and used for the web API earlier in the book. To do this, Sarah needs the cloud flow's URL. This can be found in Power Pages Studio, as shown in *Figure 10.12*:

Figure 10.12 – Cloud flow URL

Sarah created a web template called `Account` earlier and can now access the code from Power Pages Studio by selecting the code option on the client page.

Sarah has worked out that the Microsoft example is quite easy to modify. She decides to modify the following code snippets:

- `var companyNumber`

- `var _url`

- The resulting block for setting the four address fields – that is, `address line1`, `address line 2`, and `City`, and `PostCode`

Here's Sarah's final code:

```
var companyNumber = $("accountnumber").val().trim(); // Retrieve and
trim the company number input value
var _url = "https://xxxxx.powerappsportals.com/_api/cloudflow/v1.0/
trigger/1b169a46-deff-c905-fe3d-d0bba208d4a1"; // Define the URL for
the Cloud Flow trigger endpoint
var data = {}; // Create an empty object to store data
data["company number"] = companyNumber; // Set the company number in
the data object
var payload = {}; // Create an empty object for the payload
payload.eventData = JSON.stringify(data); // Convert the data object
to a JSON string and set it as the eventData property of the payload
object
shell.ajaxSafePost({ // Make a POST request using the shell.
ajaxSafePost method
    type: "POST", // Specify the HTTP method as POST
    contentType: "application/json", // Set the content type of the
request
    url: _url, // Set the URL for the request
    data: JSON.stringify(payload), // Convert the payload object to a
JSON string and send it as the request body
    processData: false, // Prevent jQuery from automatically
processing the data
    global: false, // Disable global AJAX event handlers
})
.done(function (response) { // Handle the successful response
    const result = JSON.parse(response); // Parse the JSON response
into an object
    $("address1_line1").val(result["AddressLine1"]); // Set the value
of the address line 1 input field
    $("address1_line2").val(result["AddresLine2"]); // Set the value
of the address line 2 input field
    $("address1_city").val(result["City"]); // Set the value of the
city input field
```

```
    $("address1_postalcode").val(result["Zip/PostCode"]); // Set the
value of the postal code input field
})
.fail(function () { // Handle the failure of the request
    alert("failed"); // Show an alert indicating failure
});
```

Let's take a closer look:

1. First, the code retrieves the company number input value and trims any leading or trailing whitespace.

2. Next, it defines the URL for the cloud flow trigger endpoint.

3. Then, it creates an empty object to store data and sets the company number in it.

4. After, it creates an empty object for the payload and sets the data object as the `eventData` property of the payload after converting it into a JSON string.

5. Next, it makes a POST request to the cloud flow trigger endpoint using `shell.ajaxSafePost`.

6. Then, it handles the successful response by parsing the JSON response and setting the values of address page fields based on the response data.

7. Finally, it handles the failure of the request by showing an alert message.

Sarah thought that if she could master this pattern, there were so many places where it could be used to provide a better user experience and also to do work where experienced plugin programmers were needed prior.

Sarah's journey into Power Automate was fueled by her discovery of Power Pages connectors, offering rapid data retrieval for page usage. Exploring the platform's potential through client discussions and learning materials, she recognized the importance of practical experience. Sarah envisioned implementing notification flows triggered by record changes, such as status updates, and embedding flows directly into Power Pages for executing processes and dynamic page rendering. This innovative approach aimed to streamline processes previously requiring complex JavaScript or Dataverse plugins.

The key features of Power Automate, including its workflow designer, connectors, custom actions, and data transformations, provided Sarah with the flexibility to create intricate workflows tailored to specific business needs. As she delved deeper into Power Automate, Sarah understood its pivotal role within Power Pages, enabling advanced automation scenarios and seamless integration with external systems.

Summary

Sarah's hands-on experience involved collaborating with her client to establish a comprehensive timesheet system within Dataverse, leveraging cloud flows for automated email notifications upon timesheets requiring approval. This pattern of automation not only augmented communication channels but also laid the foundation for fulfilling broader notification requirements.

Sarah's design aimed to seamlessly integrate Power Automate cloud flows within Power Pages to automate processes such as retrieving accurate company addresses for credit checks. By configuring cloud flows and integrating them into JavaScript within the web template, Sarah could trigger flows from page changes and display response results on the page, enhancing efficiency and accuracy in credit check processes.

Her implementation involved creating a cloud flow to handle HTTP requests for retrieving company addresses, configuring it within Power Pages, and integrating it into the web template's JavaScript. This approach empowered Sarah to automate address retrieval within Power Pages, demonstrating her proactive approach to mastering the platform and meeting her client's needs effectively.

In the next chapter, Sarah will look at other user experience areas – specifically dashboards and charts – that can improve pages with visualizations.

11

Charts, Dashboards, and Power BI

In this chapter, Sarah embarks on a journey to understand and utilize data-driven visualizations, specifically charts, dashboards, and Power BI, within the context of **Power Pages**. These tools are pivotal in transforming complex data into visually appealing and interactive formats, aiding in insightful decision making.

In today's rapidly evolving business landscape, data plays a pivotal role in driving informed decision making and achieving organizational success. As businesses generate vast amounts of data from various sources, the challenge lies in effectively presenting this information to key stakeholders in a way that is easily digestible, actionable, and insightful. This is where data-driven visualizations, such as charts, dashboards, and Power BI, come into play as powerful tools for transforming raw data into meaningful insights.

Charts, dashboards, and Power BI are integral components of data visualization within the context of Power Pages. These tools enable organizations to present complex data in a visually appealing and interactive manner, making it easier for users to comprehend and analyze critical information. With the ability to create dynamic visualizations, users can explore data from different angles, uncover patterns, and identify trends, ultimately leading to data-driven decisions that drive business growth and efficiency.

Charts serve as a concise and informative way to represent data through various graphical representations, such as bar charts, line charts, and pie charts. Dashboards, on the other hand, aggregate multiple charts and visual elements onto a single interface, providing a holistic view of business performance and KPIs. By embedding Microsoft Dataverse charts in Power Pages, organizations can seamlessly integrate data from their Dataverse systems, further enriching their data-driven visualizations.

Additionally, Power BI, a robust and widely used business intelligence platform, offers advanced capabilities for creating interactive and insightful dashboards. By incorporating Power BI charts and dashboards within Power Pages web templates, organizations can leverage the full potential of Power BI to deliver data-driven experiences to their portal users.

The significance of data-driven visualizations in Power Pages lies not only in their ability to present data effectively but also in their potential to empower users. These visualizations allow users to interact with the data, filter information, and drill down into specific details, granting them the flexibility to explore and extract valuable insights independently.

In this chapter, we will explore the world of data-driven visualizations in Power Pages. We will delve into the concepts and functionalities of charts, dashboards, and Power BI integration, providing a comprehensive understanding of their role in presenting data-driven insights.

Sarah explores the following topics in this chapter:

- Understanding charts and dashboards
- Embedding Microsoft Dataverse charts in power pages
- Leveraging liquid chart tags for custom web templates
- Retrieving data for charts
- Designing visually appealing charts and dashboards
- Creating interactive dashboards with Power BI
- Customizing pages for Power BI dashboards
- Optimizing chart and dashboard performance
- Real-world examples and case studies

By the end of this chapter, Sarah will be equipped with the knowledge and tools to create visually engaging and interactive data visualizations that empower her users to make data-driven decisions with confidence.

Understanding charts and dashboards

Sarah will learn that charts and dashboards are key to effective data visualization. Charts, in various forms, such as bar, line, or pie charts, provide a visual representation of data, simplifying complex information. Charts and dashboards are essential components of data visualization that play a crucial role in presenting complex data in a clear and concise manner. They enable users to understand data patterns, trends, and relationships quickly. Charts are graphical representations of data that visually depict information, while dashboards provide a consolidated view of multiple charts and data elements onto a single interface.

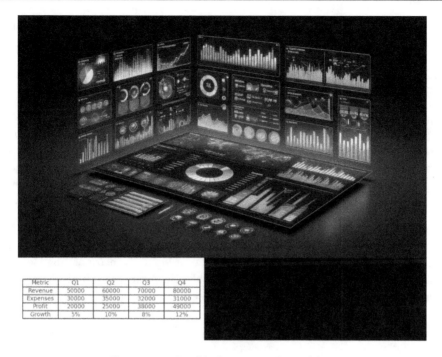

Metric	Q1	Q2	Q3	Q4
Revenue	50000	60000	70000	80000
Expenses	30000	35000	32000	31000
Profit	20000	25000	38000	49000
Growth	5%	10%	8%	12%

Figure 11.1 – Graphical representations of data

Exploring chart types and use cases

There are various types of charts available, each serving a specific purpose and offering unique insights into the data. Common chart types include bar charts, line charts, pie charts, area charts, and scatter plots. Bar charts are effective for comparing categorical data, line charts depict trends over time, pie charts show proportions and percentages, and scatter plots reveal relationships between variables. Understanding the characteristics and appropriate use cases for each chart type allows users to choose the most suitable visualization for their data.

The role of dashboards in business performance

Dashboards provide a centralized and comprehensive view of **key performance indicators (KPIs)** and critical metrics, enabling users to monitor business performance in real time. By aggregating data from multiple sources, dashboards present an overview of various aspects of the business, such as sales, marketing, finance, and customer service. They provide a holistic perspective, allowing users to identify trends, track progress, and make data-driven decisions. Dashboards facilitate efficient data analysis and enable stakeholders to stay informed and take proactive actions to drive business success.

In Power Pages, understanding charts and dashboards is crucial for creating impactful data visualizations that effectively communicate insights to users. By leveraging different chart types and designing intuitive dashboards, organizations can enhance their data-driven decision-making processes, improve operational efficiency, and gain a competitive edge. In the following sections, Sarah will delve deeper into the intricacies of creating and customizing charts and dashboards in Power Pages, empowering her to unlock the full potential of data visualization in her client's organization.

Leveraging Liquid chart tags for custom web templates

Liquid chart tags in Power Pages enable Sarah to create dynamic, interactive charts tailored to specific needs. She follows a process of defining data sources, integrating Liquid tags for chart rendering, customizing chart appearance, and adding interactivity. This process allows Sarah to create engaging and informative visualizations, enhancing the user experience on her Power Pages.

Liquid chart tags offer a powerful solution for displaying dynamic data visualizations within custom web templates in Power Pages. As a templating language, Liquid enables developers to create flexible and interactive charts that adapt to changing data and user input. By incorporating Liquid chart tags, organizations can enhance their web templates with visually appealing and informative data visualizations that provide real-time insights to portal users.

Demonstrating how to create custom web templates

Creating custom web templates with Liquid chart tags is a straightforward process that allows developers to showcase interactive data visualizations tailored to their specific requirements. In the following subsections, we'll see a demonstration of how to create custom web templates using Liquid chart tags.

Sarah produces an Agile user story to help her approach a basic chart to learn how to use Liquid chart tags.

Agile user story – creating custom web templates with Liquid chart tags

Title: Visualizing Contact Revenue with Custom Web Templates

As a: Web Developer

I want: to create custom web templates using Liquid chart tags

So that: I can showcase interactive data visualizations tailored to specific business requirements.

Acceptance criteria

Here are the acceptance criteria:

- **Define data source**: Identify the data source from which the chart will retrieve dynamic information. This data can come from various sources, such as Dataverse entities, custom data sources, or external APIs.

- **Integrate Liquid tags**: Integrate Liquid chart tags into the web template's code to dynamically render charts based on the data source. Liquid tags enable developers to fetch and manipulate data to create a wide range of charts, including bar charts, line charts, and pie charts.

- **Customize chart appearance**: Customize the appearance of the charts using Liquid tags to match the website's branding and user interface. Developers can adjust chart colors, fonts, and styles to align with the overall design of the web template.

- **Implement interactivity**: Leverage Liquid chart tags to add interactivity to the charts, allowing users to interact with the data and explore different aspects of the visualizations. Interactive features such as tooltips, data filtering, and drill-down options enhance the user experience.

Step 1 – define the data source

Identify the data source from which the chart will retrieve dynamic information. This data can come from various sources, such as Dataverse entities, custom data sources, or external APIs.

Step 2 – integrate Liquid tags

Integrate Liquid chart tags into the web template's code to dynamically render charts based on the data source. Liquid tags enable developers to fetch and manipulate data to create a wide range of charts, including bar charts, line charts, and pie charts.

Step 3 – customize chart appearance

Customize the appearance of the charts using Liquid tags to match the website's branding and user interface. Developers can adjust chart colors, fonts, and styles to align with the overall design of the web template.

Step 4 – implement interactivity

Leverage Liquid chart tags to add interactivity to the charts, allowing users to interact with the data and explore different aspects of the visualizations. Interactive features such as tooltips, data filtering, and drill-down options enhance the user experience.

To illustrate the practical application of the steps for creating custom web templates with Liquid chart tags, let's delve into a specific code example. This example will demonstrate how the Liquid chart tags can be effectively used to fetch data, render an interactive chart, and customize its appearance. By examining this code snippet, Sarah gains a clearer understanding of how each element of the Liquid chart tag works in harmony to create a dynamic and visually appealing data visualization. The following example is crafted to demonstrate not only the structure but also the versatility and the potency of Liquid in transformation.

Example – Liquid chart tags code

The following example is designed to showcase not only the structure but also the versatility and the power of Liquid in transforming raw data into engaging and informative charts within Power Pages:

```
{% chart
  data: portal.contacts
  type: 'bar'
  x: 'name'
  y: 'revenue'
  title: 'Top Contacts by Revenue'
  options: {
    scales: {
      xAxes: [{
        ticks: {
          autoSkip: false,
        },
      }],
    },
  }
%}
```

Let's look at the preceding example in detail:

- `{% chart %}`: This is the Liquid chart tag that indicates the beginning of the chart block. It allows Sarah to create and customize a chart within the web template.
- `data: portal.contacts`: This specifies the data source for the chart. In this example, the chart will use data from the contacts entity in Power Pages.
- `type: 'bar'`: This defines the type of chart to be displayed, which is a bar chart in this case. Other supported chart types include `'line'`, `'pie'`, `'doughnut'`, and `'radar'`.
- `x: 'name'`: This indicates the data field that will be used for the x-axis labels. In this example, the chart will use the `'name'` field from the contacts entity to label the bars on the x-axis.
- `y: 'revenue'`: This specifies the data field that will be used for the y-axis values. In this case, the `'revenue'` field from the contacts entity will determine the height of each bar on the chart.
- `title: 'Top Contacts by Revenue'`: This provides a title for the chart, which will be displayed above the chart to describe its content.

- `options: { ... }`: This section allows the customization of various options and settings for the chart. In this example, the `options` object is used to configure the x-axis behavior, specifically setting `autoSkip: false` to display all x-axis labels without skipping any.

The preceding code snippet creates a bar chart using Liquid chart tags in a Power Pages web template. The chart will display the top contacts by revenue, where each bar represents a contact's revenue, and the x-axis labels show the names of the contacts. The chart is highly customizable to suit specific business requirements, such as adjusting colors, adding tooltips, and enabling interactivity.

Test and optimize

Thoroughly test the custom web templates with Liquid chart tags to ensure that the charts function as intended and respond to various scenarios. Optimize the performance and responsiveness of the charts to deliver a seamless user experience.

The flexibility of Liquid chart tags

One of the key advantages of using Liquid chart tags is their flexibility in rendering dynamic and personalized charts based on user input. Liquid tags allow developers to create conditional logic and fetch data from multiple sources, enabling the creation of charts that adapt to user interactions and preferences.

For instance, developers can create charts that change based on the selected date range, location, or other parameters. This flexibility empowers users to customize their data visualizations and gain deeper insights into the information presented.

Moreover, Liquid chart tags can be combined with other Liquid features, such as loops and variables, to further enhance the complexity and richness of the data visualizations. This versatility allows developers to create truly interactive and engaging charts that cater to diverse user needs.

By harnessing the power of Liquid chart tags, organizations can unlock a new level of customization and interactivity in their web templates, delivering data-driven visualizations that captivate users and empower them to make informed decisions. Whether it's displaying real-time analytics, performance metrics, or business trends, Liquid chart tags offer a robust solution for creating compelling and dynamic data visualizations within Power Pages.

In the next section, Sarah will walk through how to embed a Dataverse chart into Power Pages.

Embedding Microsoft Dataverse charts in Power Pages

Sarah explores embedding Microsoft Dataverse charts into Power Pages, providing a streamlined process for visualizing Dataverse data within her website. She learns to identify chart requirements, access chart data, configure web templates, and customize chart appearances, creating a seamless integration that enriches user experience with valuable insights.

Power Pages offers seamless integration with Microsoft Dataverse, enabling organizations to leverage the rich data visualization capabilities of Dataverse charts within Power Pages. This integration allows users to access and interact with Dataverse data directly from the Power Pages website, providing a unified experience and enhancing the decision-making process. By embedding Dataverse charts in Power Pages, businesses can create compelling visualizations that offer valuable insights to website users.

Sarah authors an Agile user story to share with her client for this work.

Agile user story – embedding Microsoft Dataverse charts in Power Pages

Title: Integrating Dataverse Charts into Power Pages for Enhanced Data Visualization

As a: Web Developer

I want: to embed Microsoft Dataverse charts into Power Pages

So that: users can interact with and gain insights from dynamic data visualizations directly on the website.

Acceptance criteria

Here are the acceptance criteria:

1. **Identify chart requirements**: Determine the specific Dataverse charts that need to be embedded in the page, understanding the data and insights to be displayed to users.

2. **Access chart data**: Ensure the Dataverse data is accessible and available for integration with the portal, verifying data permissions and access settings for a secure connection.

3. **Configure web templates**: Create or modify web templates in Power Pages to accommodate the Dataverse charts, customizing them to display the charts in desired locations within the portal.

4. **Add chart components**: Utilize Liquid chart tags within the web templates to dynamically embed the Dataverse charts, fetching data from Dataverse entities based on specific criteria.

5. **Customize chart appearance**: Adjust the appearance of the embedded Dataverse charts to align with the portal's branding and user interface, enhancing visual appeal and clarity.

Step-by-step guide on embedding Microsoft Dataverse Charts

Creating Microsoft Dataverse charts using drag-and-drop chart tools is easy and quick, and there is no additional cost to implementing these charts or embedding them on a web page. Embedding Microsoft Dataverse charts in Power Pages web templates is a straightforward process that involves a series of steps to ensure a smooth integration. Here's a step-by-step guide to embedding Dataverse charts.

Step 1 – identify chart requirements

Determine the specific Dataverse charts to embed in the page. Understand the data and insights to display to users through these charts.

Step 2 – access chart data

Ensure that the Dataverse data to visualize is accessible and available for integration with the portal. Verify data permissions and access settings to ensure a secure data connection.

Step 3 – configure web templates

In the Power Pages environment, create or modify web templates to accommodate the Dataverse charts. Customize the web templates to display the charts in the desired locations within the portal's pages.

Step 4 – add chart components

Utilize Liquid chart tags within the web templates to embed the Dataverse charts dynamically. These tags allow the chart to fetch data from Dataverse entities and generate charts based on specific criteria.

Step 5 – customize chart appearance

Customize the appearance of the embedded Dataverse charts to align with the portal's branding and user interface. Adjust chart colors, labels, and formatting to enhance visual appeal and clarity.

Here is an example of embedding charts for timesheet data from Dataverse. It is recommended to do this work within a solution, following these steps:

1. Open `https://make.powerpages.microsoft.com/`.
2. Select the **Solutions** tab and open the **Buildapp** solution.
3. The incidents table already exists there from the work in *Chapter 5*.
4. Select the **Chart** tab in the incidents.
5. Select the **New Chart** button, label the chart as `Incidents by Employee`, and set the parameters as shown in *Figure 11.2*:

   A. Add a legend entry of **Name** with a `Count:All` aggregation.

   B. Add a **Horizontal** series of **Employee**.

6. Save it and make a note of the chart ID in the URL of the chart.
7. Decide which incident view to use, in this case, **All incidents**. Open the view in the solution and make a note of the view ID.
8. Open the web template to embed this chart; for example, a dashboard web template or a web page.
9. Paste the example code into the web template of the dashboard.

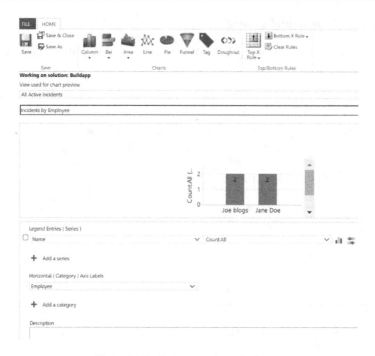

Figure 11.2 – Dataverse chart designer

Web template code for embedding a Dataverse chart

Here is the web template code for the incident dashboard showing incidents per employee:

```
<div class="dashboard container">
<div class="page-header">
<h1 aria-label="{{ snippets['Incident Dashboard Title'] | default:
resx['Incident Dashboard_Title'] | h }}">{% editable snippets
"Incident Dashboard" default: resx["Incident_Dashboard_Title"] type:
'text' tag: 'span' %}</h1>
</div>
 <div class="row">
     <div id="chartContainer">
       <div  class="col-md-12 show-chart-legend">
         {% chart id:"ba15e92c-74b4-ee11-a569-000d3a21a88f"
viewid:"fc123f69-74b4-ee11-a569-000d3a21a88f" %}
       </div>
     </div>
  </div>
</div>
```

The provided code is a snippet of a web template called "Dashboard" that includes multiple Liquid chart tags. Here's a brief explanation of the code:

- `<div class="dashboard container">`: This is a div container that defines the layout for the dashboard.

- `<div class="page-header">`: This div represents the header section of the dashboard.

- `<h1 aria-label="...">`: This is a heading tag used to display the title of the dashboard.

- `{% editable snippets "Dashboard Title" default: resx["Incident_ Dashboard"] type: 'text' tag: 'span' %}`: This is a Liquid editable tag that allows the title to be edited within the portal. The snippets object retrieves the value of the `'Dashboard Title'` snippet, and if it doesn't exist, the default `resx["Incident_ Dashboard_Title "]` value is used. The `type: 'text'` specifies that the input field should be of type text, and `tag: 'span'` indicates that the editable field should be rendered as a span element.

- `<div class="row">`: This div represents a row within the dashboard layout.

- `<div id="chartContainer">`: This is a container div that wraps the chart elements.

- `<div class=" col-md-12">`: This div represents a column in the layout, and the `col-md-12` class specifies that the column should occupy 12 out of 12 columns in medium-sized devices.

- `{% chart id:" ba15e92c-74b4-ee11-a569-000d3a21a88f" viewid:" fc123f69-74b4-ee11-a569-000d3a21a88f" %}`: This is a Liquid chart tag that generates a chart within the specified div. The `id` attribute represents the chart's unique identifier, and the `viewid` attribute represents the view (visualization) associated with the chart.

The provided code snippet demonstrates the structure of a web template for a dashboard, including the header, title, and multiple columns for displaying charts using Liquid chart tags. Each chart is associated with a specific identifier (`id`) and view (`viewid`) to determine its content. The actual chart data and customization options would be defined elsewhere in the code or through data binding.

By following this guide and leveraging the configuration and customization options available in Power Pages, organizations can effortlessly embed Microsoft Dataverse charts, enhance data visualization capabilities, and deliver valuable insights to users within their portal environment. The integration of Dataverse charts with Power Pages empowers businesses to create compelling and interactive data visualizations that drive better decision making and improve overall user engagement.

> **Tip**
>
> For further reading on how to create Dataverse charts, follow this link: `https://learn. microsoft.com/en-us/power-apps/maker/model-driven-apps/create- edit-system-chart`.

Test chart functionality

Test the embedded Dataverse charts within Power Pages to ensure their accuracy, responsiveness, and functionality. Verify that users can interact with the charts and access relevant data. Having created the example incidents chart, Sarah can test it by viewing the page on the website, while examining the data to ensure the visualization is an accurate reflection of the data, as shown in *Figure 11.3*.

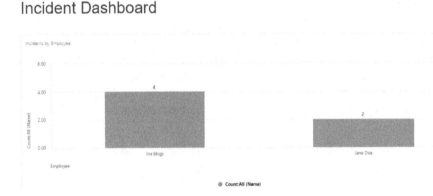

Figure 11.3 – Rendered chart on web Page

In the next section, we will learn how to retrieve data for charts.

Data for charts

To create meaningful and impactful charts, it's crucial to retrieve data from various sources and ensure its accuracy and real-time updates. In this section, we will explore different data retrieval methods for charts within Power Pages and discuss best practices for data preparation and organization.

Understanding different data retrieval methods

Here are some of the data retrieval methods:

- **Microsoft Dataverse entities:** Power Pages seamlessly integrate with Microsoft Dataverse entities, allowing charts to fetch data directly from Dataverse tables and fields.

- **Custom data sources:** In addition to Dataverse entities, Power Pages supports custom data sources. Charts can connect to external databases, data warehouses, or other systems to retrieve data. This flexibility enables charts to incorporate data from multiple sources and create comprehensive visualizations.

- **External APIs**: Power Pages also enables integration with external APIs, allowing the retrieval of real-time data from external systems or services. This is particularly useful when up-to-date information is needed, such as stock prices, weather data, or social media metrics.

Best practices for data preparation and organization

Here are some of the best practices for data preparation and organization:

- **Data cleansing and transformation**: Before feeding the data into charts, it's essential to perform data cleansing and transformation. This includes handling missing or invalid data, normalizing data formats, and aggregating data if necessary. Clean and well-organized data ensures accurate and reliable chart visualizations.

- **Data indexing and optimization**: For large datasets, consider indexing the data fields used in charts to improve query performance. This can significantly speed up data retrieval and enhance the responsiveness of charts. Additionally, it can optimize data retrieval queries to fetch only the necessary data, minimizing processing overhead.

- **Data security and privacy**: Ensure that any sensitive or confidential data used in charts adheres to security and privacy guidelines. Apply appropriate access controls and encryption measures to protect sensitive information and comply with data protection regulations.

Ensuring data accuracy and real-time updates

Here are some ways to ensure data accuracy and real-time updates:

- **Data refresh frequency**: Determine the appropriate refresh frequency for the chart data based on its volatility and relevance. Real-time charts may require more frequent data updates, while others can be refreshed at regular intervals. Align the data refresh rate with the needs of the users and the timeliness of the insights they require.

- **Data validation and error handling**: Implement data validation techniques to ensure the accuracy and integrity of the retrieved data. Perform data quality checks, handle any errors or exceptions gracefully, and provide meaningful error messages to users when data retrieval issues occur.

- **Event-driven updates**: Utilize event-driven mechanisms or triggers to update chart data automatically when relevant events occur. For example, if a Dataverse record is updated or a new data point becomes available, trigger a data refresh to reflect the changes in real-time visualizations.

These practices enable Sarah to leverage the power of data visualization to drive data-driven decision making and enhance user experiences.

Now that Sarah has explored the retrieval of data for charts, let's shift focus to the crucial aspect of designing visually appealing charts and dashboards.

Designing visually appealing charts and dashboards

Designing visually appealing charts and dashboards is essential for effectively conveying data insights and providing an engaging user experience. Sarah learned to put her charts into a solution and they are found under charts in a table. To get a chart ID, open the chart in the chart designer and copy the GUID from the URL. *Figure 11.4* shows where Sarah's timesheet charts are:

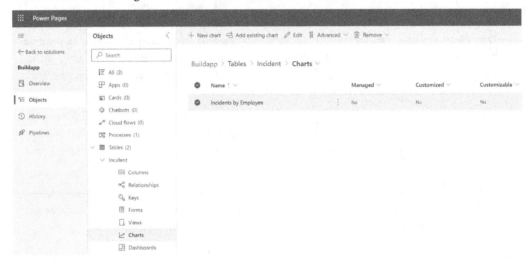

Figure 11.4 – Location of charts in a solution

Figure 11.4 shows there is a + **New chart** button where Sarah can create new charts for that table, and this opens the chart designer.

Let's look at some key considerations and guidelines to keep in mind.

Color schemes

Choose a color scheme that is visually pleasing and enhances the readability of the data. Use contrasting colors to differentiate data points and highlight important information. Consider the use of color gradients or shades to represent data trends or levels.

Data labeling

Clearly label data points, axes, and legends to provide context and make the information easily understandable. Utilize appropriate fonts, sizes, and styles for labels to ensure readability. Consider using tooltips or callouts to provide additional information or details when users interact with the charts.

Interactivity

Incorporate interactive elements to enhance the user experience and enable deeper exploration of the data. Implement features such as drill-downs, where users can navigate from summary charts to more detailed views. Include filtering options to allow users to customize the data displayed based on their preferences.

Data visualization techniques

Choose the appropriate chart type based on the data being presented and the intended message. Bar charts are effective for comparing values, line charts for showing trends over time, and pie charts for illustrating proportions. Experiment with different visualization techniques to effectively represent the data and communicate the desired insights.

Data density

Avoid overcrowding charts with excessive data points, as it can make the visualization cluttered and difficult to interpret. Use data aggregation techniques, such as grouping data into categories or intervals, to present a clearer and more concise representation of the information.

Consistency and alignment

Maintain consistency in the design elements across charts and dashboards to create a cohesive and unified visual experience. Align charts and related elements to create a logical flow and make it easier for users to navigate through the dashboard.

User feedback and testing

Gather user feedback and conduct testing to ensure that the charts and dashboards are intuitive and effectively convey the intended information. Incorporate user suggestions and make necessary adjustments to improve the overall design and usability.

By following these guidelines, Sarah can create visually appealing charts and dashboards that effectively communicate data insights and provide an interactive and engaging user experience. Remember to continuously iterate and refine the design based on user feedback and evolving business needs.

Having discussed chart design and embedding Dataverse charts, in the next section, we are going to read an overview of Power BI and its capabilities for creating interactive and insightful dashboards.

Creating interactive dashboards with Power BI

Sarah discovers the power of integrating Power BI with Power Pages, creating interactive dashboards that offer deeper insights into data. This integration enables users to interact directly with the data, offering a dynamic, enriched experience within the familiar interface of Power Pages.

Interactive dashboards created with Power BI within Power Pages represent an advancement in data visualization and user interaction. These dashboards offer more than just static reports; they provide an experience where users can delve deep into the data, uncovering trends and insights that might otherwise remain hidden. The integration of Power BI into Power Pages is not just about displaying data, but about creating an interactive environment where information becomes more accessible and meaningful.

The ability to embed Power BI charts and dashboards directly into Power Pages brings a new dimension of functionality. Users can interact with data in real time, apply filters, drill down into specifics, and even manipulate datasets within the confines of their security permissions. This level of interaction fosters a deeper understanding of the data, aiding in more informed decision making.

Furthermore, the seamless integration of Power BI into Power Pages ensures a cohesive user experience. Users don't have to switch between applications to access reports and dashboards. Instead, they can find everything they need within the familiar interface of Power Pages. This integration not only enhances user convenience but also promotes user adoption and engagement.

The step-by-step guide for embedding Power BI charts and dashboards into Power Pages ensures that even those with limited technical knowledge can successfully integrate these powerful tools. From setting up Power BI workspaces to managing access permissions, each step is designed to facilitate a smooth and secure integration process.

By meeting the necessary prerequisites, such as having the appropriate Power BI licenses and establishing data connections, organizations can unlock the full potential of Power BI within Power Pages. The result is a dynamic, data-driven portal where information is not just presented but interacted with, leading to more insightful and impactful outcomes.

Embedding Power BI charts and dashboards

Power Pages allows seamless integration with Power BI, enabling Sarah to embed Power BI charts and dashboards within her web templates. This integration allows portal users to access and interact with Power BI visualizations directly within the portal interface.

In the next section, the focus on customizing pages for Power BI dashboards highlights the adaptability of Power Pages.

Step-by-Step Guide to embed Power BI charts and dashboards

To embed Power BI charts and dashboards, Sarah follows a step-by-step process. This includes configuring Power BI workspaces, generating embed tokens, and setting up security and access permissions. The guide will walk through each of these steps:

1. **Configure Power BI workspaces**: Set up a Power BI workspace. This is where to create and store Power BI reports and dashboards.

2. **Create Power BI reports/dashboards**: In the Power BI workspace, design the reports or dashboards to embed in Power Pages.

3. **Generate embed tokens**: For each report or dashboard, generate an embed token. This token is essential for embedding and securely displaying Power BI content on Power Pages.

4. **Set up access permissions**: Ensure appropriate access permissions are in place. This step involves configuring who can view or interact with the embedded Power BI content in Power Pages.

5. **Embed in Power Pages**:

   A. **Embed in Power Pages**: Utilize the embed tokens to integrate the Power BI reports or dashboards into Power Pages. This can be done by inserting the embed codes or tokens into the relevant sections of web templates.

   B. **Customize and test**: After embedding, customize the appearance and settings as needed to ensure they align with the rest of Power Pages. Test the embedded reports or dashboards to make sure they are functioning correctly and displaying data as intended.

   C. **Manage and update**: Regularly manage and update the Power BI content as needed. This includes refreshing data, modifying reports, and ensuring the embed tokens are up to date.

Requirements and prerequisites

Before embedding Power BI charts, you'll need to ensure that you have the necessary prerequisites in place. This includes having a Power BI Pro or Premium license, creating Power BI workspaces, and configuring data connections to the relevant data sources. You'll also need to manage security and access permissions to control who can view and interact with the embedded dashboards.

Embedding Power BI charts and dashboards within Power Pages enhances the data visualization capabilities and empowers users with interactive insights. With a clear understanding of the process and the necessary prerequisites, Sarah can leverage the full potential of Power BI to create dynamic and informative dashboards that drive data-driven decision making within her organization.

Tip

Further reading on setting up Power BI integration can be found at the following link: `https://learn.microsoft.com/en-us/power-pages/admin/set-up-power-bi-integration`.

In the next section, Sarah will learn how to customize pages for Power BI dashboards and how it allows her to create tailored experiences within Power Pages, enabling users to interact with and explore data-rich Power BI dashboards. Here's how Sarah can leverage Liquid and web templates to achieve this.

Customizing pages for Power BI dashboards

To create custom pages for Power BI dashboards within Power Pages, Sarah can utilize the flexibility of Liquid and web templates. This allows Sarah to design pages that seamlessly integrate Power BI visualizations and provide a cohesive user experience. The following sections will cover some examples of Liquid code that can be used to customize pages for Power BI dashboards.

Embedding Power BI report

To embed a Power BI report on a custom page, use the following Liquid code:

```
{% powerbi_report embed_url:"https://app.powerbi.com/
reportEmbed?reportId=YOUR_REPORT_ID" %}
```

Replace YOUR_REPORT_ID with the actual report ID from Power BI. This code will render the Power BI report within the web template, allowing users to interact with the visualizations.

Filtering Power BI visuals

To enable filtering options on a Power BI visual, use Liquid code along with Power BI's JavaScript API. Here's an example:

```
<div id="powerbi-visual" style="width: 100%; height: 500px;"></div>
<script>
  var visualContainer = document.getElementById('powerbi-visual');
  var config = {
    type: 'report',
    embedUrl: 'https://app.powerbi.com/
reportEmbed?reportId=YOUR_REPORT_ID',
    ...
    // Additional configuration options
  };
  var report = powerbi.embed(visualContainer, config);
  report.on("loaded", function() {
    report.getFilters()
      .then(function(filters) {
        // Code to handle the retrieved filters
      })
      .catch(function(error) {
        // Error handling
      });
  });
</script>
```

This code embeds a Power BI visual and uses the Power BI JavaScript API to retrieve and handle filters applied by users.

- **Create a container**: A div element with the `powerbi-visual` ID acts as a container for the Power BI visual. It has a set width and height.

- **JavaScript configuration**: A script is used to configure and embed the Power BI report into the container.

- `visualContainer`: A variable that references the div container.

- `config`: An object that contains configuration settings for the Power BI report, including the type of visual (a report in this case) and the embed URL of the report.

- `report`: A variable that calls the `powerbi.embed` function to embed the Power BI report in the container using the specified configuration.

- **Event handling**: The script listens for the `'loaded'` event on the report object, which triggers once the report is fully loaded.

- `report.on("loaded", function() {...})`: This part of the script executes when the report is loaded.

- `report.getFilters()`: This retrieves the current filters applied to the report.

- `.then(function(filters) {...})`: A promise that executes after successfully retrieving filters, where Sarah can add code to handle or manipulate these filters.

- `.catch(function(error) {...})`: A catch block for error handling in case of issues while fetching filters.

Dynamic Power BI page selection – Liquid and JavaScript

Power BI provides many interactive features within dashboards. Sarah learned that apps can be built by leveraging Liquid with JavaScript, for example, with programmatic page selection on multiple pages in the dashboard by dynamically selecting a specific page based on user input:

```
<script>
  var selectedPage = '{% liquid_variable "Selected_Page" %}';
  var config = {
    type: 'dashboard',
    embedUrl: 'https://app.powerbi.com/
dashboardEmbed?dashboardId=YOUR_DASHBOARD_ID&pageName=' +
selectedPage,

    ...
    // Additional configuration options
  };
  var dashboard = powerbi.embed(visualContainer, config);
  // Rest of the JavaScript code for dashboard embedding and
interactions
</script>
```

This code retrieves the selected page name from a Liquid variable and dynamically sets the embed URL to display the corresponding Power BI dashboard page.

By using Liquid and web templates, Sarah has the flexibility to customize the layout, design, and functionality of pages that contain Power BI dashboards. This enables Sarah to create a seamless and interactive experience for users, empowering them to explore and analyze data directly within Power Pages.

Summary

In this chapter, Sarah explored the world of data-driven visualizations in Power Pages, focusing on charts, dashboards, and Power BI integration. Sarah learned that charts and dashboards are essential components of data visualization, enabling the presentation of complex data in a visually appealing and interactive manner. By embedding Microsoft Dataverse charts and leveraging Liquid chart tags, she was able to create custom web templates that showcase dynamic and personalized visualizations.

Sarah delved into different data retrieval methods, including Microsoft Dataverse entities, custom data sources, and external APIs, emphasizing the importance of data preparation and organization for optimal chart performance. She learned to design visually appealing charts and dashboards by considering elements such as color schemes, data labeling, interactivity, and data density.

Additionally, Sarah explored the capabilities of Power BI in creating interactive and insightful dashboards. By embedding Power BI charts and dashboards within Power Pages web templates, she leveraged the full potential of Power BI to deliver data-driven experiences to portal users. She also learned about customizing pages for Power BI dashboards using Liquid and web templates, enabling the creation of tailored experiences that empower users to interact with and explore data-rich visualizations.

By understanding these concepts and applying best practices, Sarah can transform raw data into actionable insights and empower users to make data-driven decisions. Data-driven visualizations in Power Pages have the potential to enhance decision-making processes, improve operational efficiency, and drive business growth. With the knowledge gained from this chapter, Sarah is now equipped to create visually engaging and interactive data visualizations that captivate users and deliver valuable insights within Power Pages.

In the next chapter, Sarah will develop integration with Xero, an accounting system using REST. This integration will enable accounts staff from her client, using Power Pages, to post batches of invoices to Xero, their accounting system.

12

REST Integration

Upon Brenda's request, Sarah, in collaboration with her client Rob the Builder, faced the task of seamlessly integrating new invoice tables managed within Dataverse. The accounts team had implemented a custom table for manual invoice entries into Dataverse. Tasked with optimizing this process, Sarah was asked to construct a web page capable of efficiently integrating and importing invoices into the accounting system. Notably, the client's accounting system operated using REST, laying the foundation for Sarah to explore and implement REST-based solutions for streamlined data exchange. Sarah, not quite knowing what would be involved, agreed and then took a few days to read and research and this is what she learned.

In the dynamic realm of Power Pages, where innovation converges with functionality, the incorporation of **REST** (short for **Representational State Transfer**) stands as a pivotal key to unlocking the full potential of seamless connectivity. For those unfamiliar, REST is not merely a term; it's a paradigm, a guiding principle shaping how applications communicate and share data on the web.

In this chapter, Sarah will implement the following:

- Design a solution to integrate the website with the Xero accounting system
- Design web pages and a basic form to support the integration
- Design a page to handle a two-step login process for REST authentication
- Design and code a web template to handle the integration and JavaScript
- Implement a Power Automate cloud flow to make HTTP requests to post invoices to the accounting system
- Implement JavaScript to call the Power Automate flow

What is REST?

At its core, REST is an architectural style defining constraints for creating web services that operate over the HTTP protocol. Picture the internet as a vast library, each book representing a piece of data or resource. REST provides a standardized way for applications to interact with these resources, akin to a library card system facilitating the retrieval of books.

One fundamental REST principle is **statelessness**. In the digital realm, each client-server request contains all necessary information, akin to handing complete instructions to a librarian. This statelessness simplifies interactions, ensuring efficiency and easy management.

Operations in REST – GET, POST, PUT, and DELETE

Delving into the language of REST, envision interacting with a smart assistant, asking it to "GET" information, "POST" new information, "PUT" updates, or "DELETE" unnecessary data. These standard operations form a universal vocabulary understood across the web:

- **GET**: Retrieve information, like asking a librarian to fetch a specific book
- **POST**: Contribute new information, adding a book to the library's collection
- **PUT**: Update existing information, akin to replacing an old book edition with a newer one
- **DELETE**: Remove information, returning a book to the librarian when no longer needed

Versatility for web services and integrations

REST's beauty lies in versatility. Without a rigid structure, it adapts to diverse application needs. In Power Pages, you can seamlessly integrate external systems, databases, and services, creating a symphony of interconnected functionalities.

Imagine Power Pages as a digital maestro, orchestrating interactions through REST's universal language. Whether retrieving real-time updates from government databases, exchanging health records, or delivering public safety alerts, REST empowers Power Pages to harmonize with diverse services.

In essence, REST integration in Power Pages acts as a universal translator for the web, allowing fluent conversations and effortless information sharing. It breaks down barriers, creates connections, and empowers users like Sarah to craft digital experiences transcending boundaries.

> **Tip**
> Further reading on REST: `https://www.restapitutorial.com/`

As Sarah delves deeper into integrating external systems with Power Pages, she discovers the transformative potential of REST for seamless data exchange and connectivity.

REST in Power Pages

Power Pages leverages REST's potential, enabling developers like Sarah to seamlessly integrate external systems, databases, and services into their projects. This integration facilitates data exchange, allowing Power Pages to interact effortlessly with diverse applications and platforms.

Companies that specialize in connectivity and integration typically provide REST **Application Programming Interfaces (APIs)** as a fundamental component of their services. A **REST API** serves as a bridge between different software applications, enabling them to communicate and share data seamlessly. Here's an overview:

Sarah's clients, spanning various sectors, could benefit significantly from REST integrations in Power Pages. In government and public service-oriented projects, REST offers unparalleled opportunities. Consider real-life examples:

- **Government databases**: Integrate Power Pages with government databases via REST for streamlined access to critical information for citizens and public servants.

- **Healthcare services**: REST integration connects Power Pages with healthcare systems, facilitating the retrieval of patient records, appointment scheduling, and access to vital health information.

- **Interoperability**: REST APIs enable interoperability between different systems and platforms. They provide a common language for diverse applications to exchange data seamlessly.

- **Scalability**: RESTful architectures are scalable, allowing companies to handle a growing volume of requests and data. This scalability is crucial for businesses experiencing expansion or increased user interactions.

- **Developer-friendly**: REST APIs are designed with simplicity and ease of use in mind. Developers can quickly understand the API structure, making integration into their applications a straightforward process.

- **Adaptability**: As technology evolves, REST APIs remain adaptable. They can be easily modified or extended to accommodate changes in business requirements or technological advancements.

Sarah discovered a common authentication and API access pattern during her exploration of Xero's API integration, specifically a two-step process with a page redirect upon login. Recognizing the ubiquity of this pattern, she expressed a keen interest in mastering its implementation to meet her client's requirements.

The identified pattern comprises a page housing a button with a URL that initiates a login page, eventually redirecting to an API redirect page. This redirect page contains an access authorization code crucial for making diverse requests, such as loading invoices, client details, remittances, or purchase orders into the accounting system. Notably, Sarah's client utilizes **Xero**, a popular hosted **SaaS** accounting application; however, the underlying principles apply universally to any REST application employing a two-step redirect pattern for access.

Figure 12.1 – REST two-step processes with redirect page

Figure 12.1 illustrates the sequence of interactions in a typical REST API authentication and access process, commonly used by applications like Xero. Here's a breakdown of the entities and their roles:

1. The user (person icon) clicks the login button
2. The login page (login screen icon) displays the authentication interface
3. The OAuth server (OAuth icon) validates the credentials and issues an authorization code
4. The redirect page (redirect icon) captures the authorization code
5. The API client (cloud icon with key) exchanges the code for an access token
6. The access token (key icon) provides secure API access
7. API requests (server icon) are made using the access token

To execute this integration, Sarah needs to configure the application within the accounting system. This involves registering her redirect page and acquiring a client ID essential for REST authentication. This setup ensures the secure and efficient flow of data between the systems, meeting the specific needs of the client's API integration requirements.

To seamlessly integrate the Xero API, Sarah formulated an agile user story to outline the necessary steps and acceptance criteria for the task.

Agile user story – Xero with Power Pages integration

Title: Implement Xero API integration for invoice submission

As a: web developer

I want: to integrate Xero's API into Power Pages to allow the automated posting of invoice batches

So that: accounts staff can efficiently submit and track invoices within the Xero accounting system

Acceptance criteria

Here are the acceptance criteria:

1. **Login page**: Create a dedicated page featuring the Xero login button
2. **Accounting app sign-in**: The login button opens the accounting app's sign-in page
3. **Redirect to custom page**: After sign-in, the app redirects to Sarah's custom redirect page
4. **Code extraction**: On the redirect page, leverage Liquid and JavaScript to extract the authorization code
5. **Power Automate flow**: Utilize Power Automate Flow to construct the API request systematically
6. **Request building**: The request generated by the Power Automate flow includes all necessary data for a secure and comprehensive POST of invoices to the Xero API

To execute this integration, Sarah needs to configure the application within the accounting system. This involves registering her redirect page and acquiring a client ID essential for REST authentication. This setup ensures the secure and efficient flow of data between the systems, meeting the specific needs of the client's API integration requirements. To ensure the seamless and secure integration of Xero's API, Sarah meticulously planned the necessary steps to configure the application within the accounting system.

Design for the Xero API integration process

Sarah formulated the following design (from the *Acceptance criteria* section above) for the Xero API integration process:

1. **Login page**: Create a dedicated page featuring the Xero login button
2. **Accounting app sign-in**: The login button opens the accounting app's sign-in page
3. **Redirect to custom page**: After sign-in, the app redirects to Sarah's custom redirect page

4. **Code extraction**: On the redirect page, leverage Liquid and JavaScript to extract the authorization code

5. **Power Automate flow**: Utilize a Power Automate flow to construct the API request systematically

6. **Request building**: The request generated by the Power Automate flow includes all necessary data for a secure and comprehensive POST of invoices to the Xero API

This design ensures the seamless flow of authentication, data extraction, and API request construction, aligning with best practices for effective Xero integration within Sarah's project.

Researching the API and Postman

Embarking on her journey to integrate with the Xero API, Sarah initiated the process with thorough research, focusing on the API documentation provided by Xero. In this exploration, she undertook the following key steps:

- **Xero API documentation study**: Sarah delved into the comprehensive Xero API documentation. This involved a meticulous examination of the available resources, guidelines, and specifications provided by Xero for developers.

- **Discovery of Postman example**: During her exploration, Sarah stumbled upon a valuable resource – a Postman example offered by Xero. This example not only included sample requests but also provided detailed instructions on how to leverage Postman for API testing and development.

- **Postman installation and usage**: To implement this newfound knowledge, Sarah installed Postman, a powerful API testing tool. Despite being a newcomer to Postman, she found well-documented guides on its installation and usage, facilitating a smooth onboarding process.

- **Hands-on testing and request building**: Using the Xero-provided Postman example, Sarah actively tested and built her initial HTTP requests. This hands-on approach allowed her to familiarize herself with the intricacies of crafting requests, understanding authentication processes, and handling data exchanges.

- **Step-by-step guide utilization**: Xero's step-by-step guide for setting up and utilizing the Postman example became a cornerstone in Sarah's learning journey. The guide covered essential aspects such as authentication POST and GET invoices examples, offering practical insights that Sarah seamlessly incorporated into her HTTP request development and testing, as shown in *Figure 12.2*.

Figure 12.2 – Postman Xero collection

Postman became Sarah's guiding light, instilling confidence and providing a clear path forward as she delved into the intricacies of developing the HTTP request. Initially encountering errors, Sarah found reassurance in the continuous functionality of the HTTP request in Postman, derived from the provided example. This iterative process allowed her to systematically identify and rectify her mistakes until, like magic, the HTTP request started functioning seamlessly. Postman, with its persistent reliability, served as both a mentor and troubleshooter throughout Sarah's journey in refining and perfecting her integration project.

By leveraging the combined power of Xero's comprehensive API documentation and the practicality of Postman, Sarah navigated the initial stages of her integration project with confidence and a solid understanding of the API functionalities.

> **Tip**
> Xero API documentation on using Postman – what Sarah studied to implement Postman: `https://developer.xero.com/documentation/sdks-and-tools/tools/postman/`

By leveraging the combined power of Xero's comprehensive API documentation and the practicality of Postman, Sarah navigated the initial stages of her integration project with confidence and a solid understanding of the API functionalities. With a solid foundation established through research and planning, Sarah proceeded to implement the Xero API integration by following a structured approach.

Implementation of Xero API integration

Sarah decided to call her redirect page xeroresponse. The Postman example included a challenge, by following the link provided in the Xero API documentation to online challenge providers. Through this link, she could input her challenge text and receive a corresponding challenge to incorporate into the authentication process.

Her strategy involved utilizing a table named batches, designed to store the records of invoices for submission. This table would not only house the invoice data but also display the status of each batch, indicating whether it was a draft or processed in Xero. This thoughtful planning ensured a systematic approach to managing and monitoring the invoice submission process.

Brenda had told Sarah that the developer in the accounts department had already set up the Xero client ID in the parent account table with a schema column name of new_xeroclientid.

Now she was ready to set up the app in Xero: she sat with her client and they logged in to the client and set up the app, which involved filling in a short form that submitted the redirect page URL that she had created earlier – xeroresponse.

POST process and designing batch submission in a Power Automate flow

Sarah developed the POST requests by studying the API example documentation and experimenting with Postman, leading her to formulate a plan. She envisioned a batch subgrid to organize batches for accounts staff, complete with a button for effortless batch posting. Leveraging Power Automate Flow's simplicity for POST requests and data preparation, she designed HTTP requests within the flow. The subsequent steps involve implementing the Power Automate flow as a child flow for flexibility, allowing manual testing, which will then be called from a parent flow with the Power Pages connection.

This approach includes manually testing the authentication process in Postman to obtain the authentication code and validate the child flow. Finally, Sarah created the trigger parent flow, initiated from the Power Pages as a triggered flow. Once Sarah had worked out her designs, she implemented the Power Automate cloud flows that would actually post the invoices.

The Power Automate child flow to post invoices

Sarah started creating a solution that she called Xero Integration. Here, she would create and store all the objects she would need to run the Xero integration. To start the Power Automate flow, Sarah created a new flow called Xero Invoice Post of the instant cloud flow type. This means it will run as a child flow when called from the Power Pages triggered parent flow she will create later.

Sarah configured this Power Automate cloud flow as follows:

1. As the first step, Sarah set three parameters – `ResponseCode`, `XeroClientId`, and `BatchId`. These will be extracted from the Xero web template on the redirect page called `XeroResponse`.

2. Before getting into the invoice post logic, Sarah needed to get the authorization code and the **Xero tenant ID**. There was an example of this both in the API documentation and also in the Postman example. She copied this HTTP request to the Get token from Postman and set an HTTP request step as shown in *Figure 12.3*, to retrieve the Xero authorization code.

3. She also copied the HTTP POST request from Postman to get Connections. This returned the Xero tenant ID, which she would need to post invoices, as shown in *Figure 12.3*.

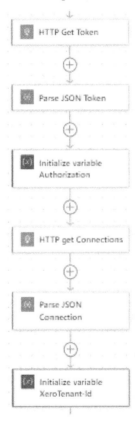

Figure 12.3 – Get token steps in the cloud flow

4. Sarah created a variable of the array type called `InvoicesItems`. Here, she will populate all the invoice details for each invoice.

5. Sarah created a Dataverse **Get row by ID** step called **Get a Batch row by ID**. This will get the batch record.

6. Sarah added a Dataverse List rows called **List Invoices Rows**, where she used a fetch XML query to get all invoices filtered by the lookup of `batchid`.

7. She added an **apply to each** loop on the invoices. The easy way was to add the next step as a Dataverse List Rows called `List Invoice Details Rows` where she used a fetch XML query to get all invoice details filtered by the invoice lookup.

    Note that when she selected the invoice from the **List Invoices Rows**, Power Automate automatically created the **apply to each** loop and embedded **List Invoice Details Rows**, as can be seen in *Figure 12.4*, when a step referenced **List Invoices Rows**.

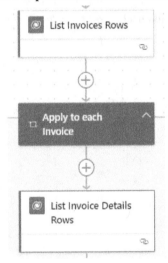

Figure 12.4 – Apply to each automatically created

8. Now she needed to fill the `InvoicesItems` array to create a line item for the invoice in the Xero Post:

    A. Sarah added a compose step to fill a description entry with meaningful information from the invoice details, which includes the Invoice Details description column.

B. Sarah added a compose step, which will complete the JSON block to include a description and other values that Xero requires. Initially, she copied this block from the API documentation and then replaced it with columns from List Invoice Details Rows, as shown in the following code block:

```
{
    «Description»: «@{outputs('ComposeDescription')}»,
    «Quantity»: «@{items('Apply_to_each_4')?['new_units']}»,
    "UnitAmount": "@{items('Apply_to_each_4')?['new_bill']}",
    "TaxType": "OUTPUT2",
    "TaxAmount": "@{items('Apply_to_each_4')?['new_tax']}",
    "LineAmount": "@{items('Apply_to_each_4')?['new_billtotal']}",
    "AccountCode": "200"
}
```

9. Next, she added a step of type Set Variable, selecting the Invoice items array. She called it Append to Invoice Items array variable, and set its value to the Compose output of item 6, as shown in *Figure 12.5*.

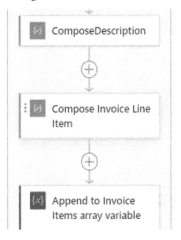

Figure 12.5 – Append invoice details to invoice Items array

10. Now Sarah had the data blocks ready to set the HTTP step for the HTTP POST request to post the invoices to Xero. For the body of the JSON block, she copied the block from the API documentation invoices page and replaced the values with the values derived from the invoice record. She added the Invoice items array variable for the JSON block needed for Line items, as shown in *Figure 12.6*.

Figure 12.6 – HTTP step to post invoices

This HTTP request is designed to post invoices to Xero's API. The headers play a crucial role in defining the request:

- **Content-Type**: Specifies that the content of the request is in JSON format
- **Accept**: Informs the server that the client expects a JSON response
- **Xero-Tenant-Id**: Identifies the Xero organization/account to which the request pertains
- **Authorization**: Contains the authentication token required for secure access to the Xero API

The body of the request includes details of the invoice to be posted, such as type, contact information, date strings, invoice number, reference, currency code, status, line amount types, subtotal, total tax, total, and line items. These details are formatted and populated based on the variables retrieved from the corresponding Power Automate flow steps. The `LineItems` field is populated with a variable called `InvoiceItems`, described in *step 8*.

This was an explanation of the key steps implemented in the cloud flow. The full cloud flow is found in the GitHub solution for Power Pages in Action, and it is worth studying each step to learn what Sarah implemented. Sarah turned the flow on and was able to test the flow by manually testing it in the flow designer and entering values for the parameters directly, which she copied from Postman. Next, she needed to create the trigger cloud flow that would run from the web page.

Adding a Power Pages triggered Power Automate cloud flow

To create a Power Pages cloud flow, Sarah opened the page editor in the Power Pages design studio, browsed to the **Set up** tab, and selected the **Cloud flows** tab, as shown in *Figure 12.8*. Sarah created a new flow by clicking the **Create new flow** button and then did the following:

1. Selected the Power Pages **When Power Pages calls a flow** trigger
2. Named the flow `Post Invoices to Xero trigger`
3. Added three input parameters: `ResponseCode`, `XeroClientId`, and `BatchId`
4. Added a new step to call the child flow by selecting **Xero Invoices Post**

Figure 12.7 – Power Pages trigger cloud flow

The parameters in Power Pages are passed through to the child flow that contains the HTTP request that uses them to connect to Xero and Post invoices. These parameters will be set in the web page web template and passed into the post.

5.   Sarah copied the URL of the cloud as shown in *Figure 12.8*.

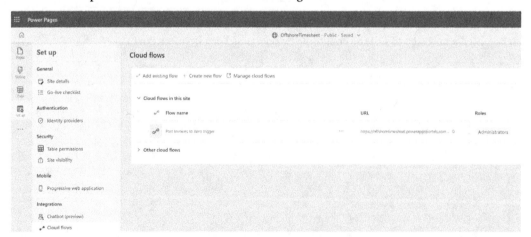

Figure 12.8 – Power Pages cloud flow

Sarah then had a working cloud flow that had been tested manually and successfully posted invoices in Xero. Sarah could then work on developing the web pages. First, Sarah would need to develop the navigation page that allows the user to log in to Xero and then develop the authentication login button and test the authentication access.

Authentication button and login page

Sarah created a web page named XeroAuthentication, featuring a simple layout with a submit button and a customized web template labeled Xero. Her plan involved employing Liquid and JavaScript to configure this button, enabling it to invoke the Xero login page, which subsequently redirects to the designated redirect page of /xeroresponse.

The code she incorporated into the web template served the purpose of configuring the submit button. It included authentication parameters essential for opening the Xero login page, a setup established during the app registration process with Xero. This strategic implementation aimed to seamlessly integrate the authentication process into the web page, ensuring a smooth user experience.

Code for the login button

Sarah designed a web template exclusively for the login page, featuring a single button. This page was strategically positioned as a sub-page within the navigation menu. Sarah foresaw the possibility of streamlining the login process further by eventually removing this button and seamlessly initiating the login directly from the navigation.

Web template code

With the design in place and a thorough understanding of the API, Sarah proceeded to implement the Xero API integration, beginning with configuring the essential application settings and constructing the necessary web templates:

```
{% block main %}
{% include 'Page Copy' %}
<div class="container" role="main">

{% if user.id %}
{% assign usercontact    = entities['contact'][ User.id] %}
{% assign parentaccountid = user.parentcustomerid.id %}
{% assign parentaccount  =  entities['account'][parentaccountid] %}
{% assign xeroclientid =  parentaccount.new_xeroclientid %}
{% endif %}

<script type="text/javascript">
$(document).ready(function() {
  var xeroClientId = "{{ xeroclientid }}";
```

```
    var updateButton = document.getElementById('UpdateButton');
        if (updateButton) {
            updateButton.onclick = function() {
                // Perform validations or any other logic needed
before redirect
                if (typeof entityFormClientValidate === 'function' &&
!entityFormClientValidate()) {
                    return false;
                }
                // Construct the URL with the xeroClientId
                var baseUrl = «https://login.xero.com/identity/
connect/authorize»;
                var queryParams = new URLSearchParams({
                    response_type: 'code',
                    code_challenge_method: 'S256',
                    client_id: xeroClientId, // Use the xeroClientId
from the Liquid variable
                    scope: 'accounting.contacts openid profile email
accounting.transactions offline_access',
                    redirect_uri: 'https://YOURWEBSITE.
powerappsportals.com/xeroresponse',
                    code_challenge: 'dl12345nNJ8vSxfthukYcCe8oX-
OFHes6765432173Y'
                });
                // Redirect to the constructed URL
                window.location.href = `${baseUrl}?${queryParams.
toString()}`;
                return false; // Prevent default form submission
            };
        }
    });
</script>
</div>
{% endblock %}
```

This code is a web template that integrates Xero's authentication into a Power Pages environment. Let's look at a brief breakdown of the key components.

Block and Include statements

{% block main %} and {% include 'Page Copy' %}: These lines define the main content block and include additional content from another template named 'Page Copy'.

Container setup

`<div class="container" role="main">`: This creates a `div` container with a main role for semantic HTML.

User data retrieval

- `{% if user.id %}`: This checks whether the user is logged in.
- `{% assign usercontact = entities['contact'][User.id] %}`: This assigns the logged-in user's contact entity.
- `{% assign parentaccountid = user.parentcustomerid.id %}`: This retrieves the ID of the user's parent account.
- `{% assign parentaccount = entities['account'][parentaccountid] %}`: This assigns the parent account entity.
- `{% assign xeroclientid = parentaccount.new_xeroclientid %}`: This retrieves the Xero client ID from the parent account entity.

JavaScript section

- `<script type="text/javascript">$(document).ready(function() {...});</script>`: This ensures the script runs when the document is fully loaded.
- `var xeroClientId = "{{ xeroclientid }}";`: This stores the Xero client ID in a JavaScript variable.
- `var updateButton = document.getElementById('UpdateButton');`: This gets the reference to the update button by its ID.
- `if (updateButton) { updateButton.onclick = function() {...}; }`: This attaches an `onclick` event handler to the button if it exists.

OnClick event handler

`if (typeof entityFormClientValidate === 'function' && !entityFormClientValidate()) { return false; }`: This performs form validation if a validation function exists.

It constructs the URL for Xero's authorization endpoint using query parameters, including the client ID, scope, and redirect URI.

`window.location.href = ${baseUrl}?${queryParams.toString()};`: This redirects the user to the constructed URL for Xero authentication.

End block

{% endblock %}: This ends the main content block.

This code facilitates secure authentication and authorization with Xero by redirecting the user to Xero's login page and handling the authentication process seamlessly within the Power Pages environment. This web template code includes 'Page Copy' content. It checks whether a user is logged in, retrieves relevant data, and assigns the Xero client ID. The JavaScript code, executed when the document is ready, handles the button click event. It constructs a URL with the Xero client ID and redirects to the Xero authorization page. The URL includes parameters such as response type, client ID, scope, and redirect URI. This code facilitates secure authentication and authorization for Xero integration in the Power Pages environment. It loads an Xero authorization page, as shown in *Figure 12.9*.

Figure 12.9 – Xero login page

After successfully testing the Xero login page from the button and confirming a successful login, Sarah was ready to develop the redirect page called XeroResponse.

XeroResponse page

This serves as the designated redirect page for Xero after the user logs in. The URL query includes the authentication code, required for making subsequent calls. The primary purpose of this page is to facilitate Xero POST requests, enabling operations on batches that the user intends to submit to Xero. The following subsections detail the various elements implemented on this page.

Web template

The web template will contain the following:

- Liquid code to assign variables for server database values
- JS code containing a POST request for each operation on the page

A list of batches that the user can inspect, manage, and process to post to Xero; JS code to deliver responses to the user from the posts indicating failure or success in posting a batch

Liquid

In the process of Xero integration, Sarah utilizes Liquid and JavaScript to extract the authentication code and retrieve the Xero Client ID, essential for subsequent POST requests to Xero, as demonstrated in the code snippet in the next section.

Sarah could assume that the user was an employee of the company, making the POST and also having an authorized role, and could therefore extract `xeroclientid` from the employer's account – in this case, from Rob the builder's account records, as it would be an accounts team employee who would be logged in.

Xero Client ID

The following code snippet demonstrates how the Xero Client ID is retrieved based on the logged-in user's information:

```
{% if user.id %}
{% assign usercontact  = entities['contact'][ User.id] %}
{% assign parentaccountid = user.parentcustomerid.id %}
{% assign parentaccount =  entities['account'][parentaccountid] %}
{% assign xeroclientid =  parentaccount.new_xeroclientid %}
{% endif %}
```

The Xero Client ID is retrieved based on the logged-in user's information. Additionally, the Xero authentication code is extracted from the URL query parameters.

Xero authentication code

The following code snippet demonstrates how the Xero authentication code is extracted from the URL query parameters:

```
{% if request.params['code']%}
{% assign authCode = request.params['code'] %}
{% endif %}
```

JavaScript

Within the JavaScript section, Sarah establishes variables set from the Liquid code in the web template as follows:

```
<script type="text/javascript">
$(document).ready(function() {
var authCode =    "{{code}}";
var xeroClientId = "{{ xeroclientid }}";
});
</script>
```

Within the document ready function, two variables (`authCode` and `xeroClientId`) are set using the Liquid variables. These JavaScript variables will be used for making subsequent POST requests to Xero.

With the implementation of the JavaScript **document ready function** variables, and subsequent POST requests to Xero, her journey into the development of the POST process, and the design of batch submission in a Power Automate flow takes shape. Finally, Sarah was ready to develop the redirect page that would call the cloud flows she had taken a couple of days to develop and test.

Xero response web page content

The Xero response web page is the redirect page in Xero that Sarah configured earlier, together with her client. This page will allow the user to select batches and post the batch of invoices using the cloud flow. In Power Pages studio, Sarah created a new page called `Xeroresponse`. Sarah started to think about how this would work in detail. She knew that she would have to call the cloud flow per batch, that is, the user would have to select a batch and then there would need to be a button or link for the user to call the cloud flow. Ideally, Sarah wanted to be able to select a batch from a subgrid of batches and then call the post invoices flow on that batch and return a success message on the page. Sarah realized that she could not use the Power Pages configuration to easily do this. In a subgrid, it is not possible to have a custom column where she could configure a button or link with its batch ID to send to the cloud flow that requires the three parameters. Sarah considered her choices.

Sarah could create a complete custom page in HTML in the web template that would list the batches with a column containing a link to post the batches in the cloud flow. The complexity of this option included a custom fetch XML query to get the batches, develop the batches table, and a column button that would pass the parameters to a function that would call the post-cloud flow.

Another option would be to have a details button on the subgrid that would open the batch in its basic form and configure a button to post the batch to the cloud flow. This seemed a bit easier, as it involved less custom code, but it would need a JavaScript event to populate the **Post Invoices** button. Sarah would need to use a placeholder for the button, as a JavaScript event button does not exist in Power Pages, and would also need to get the other parameters, namely `ResponseCode`, into the URL query of the details page. This was a challenge, as with this option the design would require that when the batch details page opened, it could access `ResponseCode`, which the post invoices cloud flow is dependent on to make the HTTP request to Xero.

Another option would be to have a list page of batches. Sarah could use JavaScript to insert an HTML table column in the list as it loads and the column could contain a link that would post that batch. This seemed like a nice way to customize the page. Sarah created a list she called `Xero Batch`. She then configured the `XeroResponse` page to the `Xero Batch` list. She set the default view of batches that were in the draft status so that users would not attempt to process batches that had already been processed. When this page was ready, she ran the process and logged into Xero. It opened the `XeroResponse` page and she opened the browser developer tools for that page and examined the names of the columns to find a way to identify them so she could insert or inject an HTML column. Sarah was ready to build the custom column for the list.

Custom column

Sarah needed to identify a column to insert a new column beside it and Sarah would need to develop a link button for that column that would post the invoice to Xero by calling the cloud flow.

Sarah examined the rendered HTML in the developer tool. She noted that the row contained the batch ID GUIDs with a property of `data-id`, which she would need to post in the cloud flow:

```
<tr data-id="4f997a4c-86b1-ee11-a81c-002248c7bc1e" data-entity="new_
batch" data-name="xxxxx">
```

Now she needed to identify a column in a unique way. She examined the first column of the list:

```
<td data-type="System.String" data-attribute="new_name" data-
value="Xero Invoice Integration" aria-readonly="true" data-th="Name"
aria-label="xxxxxxx">xxxxxxxxx</td>
```

The first column of the list was the name of the batch and it contained `data-attribute="new_name"`. This seemed unique enough to use JavaScript to identify the column.

Sarah worked out the JavaScript code and created a function in the web template called `CreatePostBatchColumn` as follows:

```
// Function to create a new column with a link button for posting
invoices
function CreatePostBatchColumn() {
    // Select all table rows in the list
    $("table.data-table tbody tr").each(function() {
        // Extract batch ID from the data-id attribute
        var batchId = $(this).data("id");
        // Find the table column with data-attribute="new_name"
        var nameColumn = $(this).find("td[data-attribute='new_
name']");
        // Create a new column with a link button
        var postColumn = $("<td>").html("<button class='post-button'
onclick='PostInvoices(\"" + batchId + "\")'>Post Invoices</button>");
        // Insert the new column beside the nameColumn
        nameColumn.after(postColumn);
    });
}
```

The `CreatePostBatchColumn` function iterates through each table row in the data table. It extracts the `data-id` attribute from the row, sets the `batchId` variable, finds the table column with `data-attribute="new_name"`, creates a new column with a link button that calls the `PostInvoices` function with `batchId` as a parameter, and inserts the new column beside the `new_name` column.

Sarah then worked on the `PostInvoices` function, which calls the cloud flow:

```
function PostInvoices(batchId) {
    // Get dynamic values from liquid variables
    var xeroClientId = "{{ xeroclientid }}";
    var authCode = "{{ code }}";

    // Define the flow URL
    var flowUrl = "https://xxxxxx.powerappsportals.com/_api/cloudflow/
v1.0/trigger/d3854766-e680-4209-1c3b-b48847c621de";
    // Prepare data for the request
    var flowData = {
        ResponseCode: authCode,
        XeroClientId: xeroClientId,
        BatchId: batchId
    };
    // Make an AJAX request to trigger the cloud flow
    shell.ajaxSafePost({
```

```
        type: «POST»,
        contentType: «application/json»,
        url: flowUrl,
        data: JSON.stringify({ «eventData»: JSON.stringify(flowData)
}),
        processData: false,
        global: false
    })
    .done(function(response) {
        // Parse the response
        var result = JSON.parse(response);
        if (result.processed === "False") {
            // Display specific error message
            alert(result.errormessage || 'Xero Process failed. Please
try again.');
        } else {
            // Processed successfully, set status to Processed
        }
    })
    .fail(function(error) {
        // Handle error here
        console.error('Error during Xero request:', error);
        alert('Error during Xero request.');
    });
}
```

Here's a detailed explanation of each point for the function provided:

- **Function purpose**: Triggers a cloud flow for processing Xero invoices

- **Dynamic values**: Retrieves a Xero client ID and authorization code from Liquid variables

- **Flow URL**: Specifies the URL for triggering the cloud flow

- **Data preparation**: Constructs data object (`flowData`) with relevant information, including batch ID

- **AJAX request**: Initiates a POST request to the specified flow URL with JSON data

- **Success handling**: Parses the response and displays an alert for success or failure

- **Error handling**: Displays an alert and logs errors if the request fails

- **Usage**: The function is designed to be called with a specific batch ID parameter

Having completed the Xero integration, Sarah reflected on the overall process and the lessons learned.

Summary

Sarah embarked on a journey to integrate Xero's API into Power Pages, demonstrating a systematic approach to RESTful architecture. She delved into Xero's API documentation, discovered valuable Postman examples, and installed Postman for hands-on testing. Sarah implemented a two-step authentication pattern, designed web pages, and coded cloud flows in Power Automate for seamless Xero integration. Leveraging Liquid and JavaScript, she crafted authentication and redirect pages, while also creating a dynamic batch submission process. Sarah faced challenges in Power Pages configuration, leading her to employ custom HTML, JavaScript, and Liquid code to enhance user interactions. Her journey showcased a nuanced understanding of API integration, Power Automate, and Power Pages customization.

In the future, Sarah could enhance the Xero response page to allow other types of batches, for example, for purchase orders and remittances. In the next chapter, Sarah will learn how to create a PDF file from Dataverse data and implement a link on a page by injecting HTML into HTML on a form.

Creating a PDF File from Dataverse

Sarah's journey into creating PDF files from Dataverse invoices begins with a pivotal request from her client.

Sarah is asked by her client to create a PDF version of an invoice record and have it so that it can be downloaded by their customers from the web page. Sarah's client explains that they want to turn data from the invoice record into a PDF that can be attached to emails and is also accessible from the web page.

Tasked with enabling customers to download PDF versions of their invoices from a web page, Sarah dives into a comprehensive exploration of design choices and implementation strategies. As she delves deeper into the project, Sarah navigates through various considerations, from the initial design phase to the seamless integration of tools and technologies. Let's explore how Sarah approaches each aspect of the process.

The following points will be covered in this chapter:

- Client requirements: Generate PDF invoices for customer accessibility and email attachments

- Design considerations: Ensure seamless PDF generation

- Utilization of Dataverse file fields: Store PDFs in Dataverse file fields, addressing challenges such as user access and file replacement

- Office Word templates: Streamline PDF generation using an Office Word template

- Power Automate cloud flows: Automate PDF generation with Power Automate

- OneDrive integration: Integrate OneDrive for converting Word templates to PDFs, considering cost-effectiveness and accessibility

- HTML layout and styling: Enhance user accessibility by injecting HTML elements

As Sarah embarks on her implementation journey, she writes an Agile user story to share with her client and ensure the successful execution of the PDF generation process.

Agile user story – create invoice PDF

Title: Create PDF invoices from Dataverse records

As a: Web developer

I want: To generate PDF files from Dataverse invoice records and enable customers to download these PDFs from the web page

So that: Customers can easily access and download their invoice copies for their records

Acceptance criteria

1. File generation:

   - Utilize an Office Word template for invoice data
   - Convert the Word template to a PDF using Power Automate
   - Store the generated PDF in the Dataverse file field

2. File accessibility:

   - Ensure the PDF file can be downloaded from the web page
   - Prevent users from replacing the generated PDF
   - Enable the download link only when the invoice status is marked as **Filled**

3. Email integration:

   - Attach the PDF to emails sent to customers
   - Implement a button on the web page to trigger sending the email with the PDF attachment

4. Implementation steps:

   - Create and configure necessary Dataverse file columns
   - Develop a Power Automate cloud flow for PDF generation and storage
   - Integrate OneDrive for Word-to-PDF conversion
   - Inject HTML into the web page for the PDF download link

With the story approved by Sarah's client, she next completes a design of the process.

Design of the invoice PDF generation process

Sarah has to consider her options in the design considerations and choices she makes to streamline the process of generating PDF files from Dataverse invoices. Each aspect of the design, from the utilization of Dataverse file fields to the implementation of a Power Automate cloud flow, plays a crucial role in achieving the desired outcome efficiently and effectively.

Design choices

Sarah carefully considers several design choices to ensure the seamless generation and management of PDF files. Each choice is made with the overarching goal of optimizing user experience, ensuring data integrity, and enabling efficient workflows.

The design has to take these points into consideration:

- The PDF file is to be generated by a process
- Users should not be able to change or replace the file
- Users can download the PDF
- Processes should be able to send the email with the PDF as an attachment on demand using a button on the page
- The invoice PDF should be available only when the invoice is processed

Sarah opts to utilize the Dataverse file field for storing the PDF generated by her process. Initially, she planned to include the file field on the page, allowing users to access and download the file. However, upon implementing the file and uploading a test PDF, she realizes a potential issue: granting users write access to the page could enable them to replace the PDF by uploading a replacement. Even worse, when the invoice record transitions to a read-only state upon approval, rendering it inactive, the file field appears on the page but is unresponsive and disabled, so users are not able to download it.

Office Word template

Sarah opts to leverage an Office Word template as the foundation for generating PDF invoices. This choice facilitates the structured representation of invoice data and allows for easy customization to meet specific client requirements. The client required it.

Her other choice is to use HTML and generate an invoice layout that way, but her client, like so many organizations, uses Office document templates for creating documents such as invoices and remittances.

Power Automate cloud flow

The decision to implement a Power Automate cloud flow for PDF generation underscores Sarah's commitment to automation and efficiency. By automating the process, Sarah streamlines the generation of PDF invoices, reducing manual effort and minimizing the risk of errors.

Conversion step with OneDrive

Integrating OneDrive into the workflow offers a seamless solution for converting populated Word templates into PDF format. This strategic decision not only guarantees the compatibility and accessibility of the generated PDF files across various platforms and devices but also provides a cost-effective solution. Sarah explores premium PDF converter connections available on Power Automate, but leveraging OneDrive proves to be an efficient and economical choice, especially considering the client's existing OneDrive storage. With just two Power Automate steps with the OneDrive connector, Sarah can create PDFs, ensuring a smooth and hassle-free process, as shown in *Figure 13.1*.

Figure 13.1 – OneDrive connector steps to create a PDF

Figure 13.1 illustrates the steps involved in creating a PDF using the OneDrive connector, demonstrating the straightforward and efficient process. Next, Sarah considers the HTML she will require.

HTML layout and styling

Appending HTML to a web page involves dynamically adding HTML elements, such as links and text, to the existing page content. In Sarah's case, she is using HTML to create a PDF link, allowing users to conveniently access and download PDF invoices directly from the web page. This approach aligns with Sarah's focus on user accessibility and intuitive interface design.

Using HTML to create the PDF link offers several benefits:

- **Enhanced user experience**: By providing a clickable link to download PDF invoices, Sarah ensures a seamless and user-friendly experience for individuals accessing the web page. Users can easily identify and interact with the link to obtain the necessary documents.

- **Flexibility and customization**: HTML allows for flexible customization of the appearance and behavior of the PDF link. Sarah can style the link to match the overall design aesthetic of the web page, enhancing its visual appeal and usability.

- **Compatibility with Power Pages**: In scenarios where it's not possible to directly add links and text to a form, such as with Power Pages, appending HTML offers a workaround solution. By injecting HTML code into the page, Sarah can dynamically insert the PDF link, overcoming limitations and providing additional functionality.

- **Scalability and maintainability**: HTML provides a scalable solution that can accommodate future changes or updates to the web page layout or content. Sarah can easily modify the HTML code to add new links or make adjustments as needed, ensuring the longevity and adaptability of the solution.

Overall, leveraging HTML to create the PDF link empowers Sarah to deliver a user-centric solution that enhances accessibility, functionality, and visual appeal, while also overcoming limitations posed by certain platform constraints such as Power Pages.

Visual flow

Sarah prepares a visual flow to further illustrate the process, as shown in *Figure 13.2*.

Figure 13.2 – Visual flow of the PDF process

Now that Sarah has considered her design choices and finalized the design, she is ready to implement the design.

Implementation of the invoice PDF process

Sarah's journey to streamline the process of generating PDF files from Dataverse invoices is detailed. The process involves creating a PDF file based on a Word template, integrating it with a Power Automate cloud flow, and injecting the PDF link into a web page. Let's dive into the steps Sarah follows to achieve this seamless workflow. First, the data model changes to support this process would need to be completed.

Dataverse table implementation

Sarah would need to create a Dataverse field to store the PDF before working on the process of creating it. Sarah, as usual, will work on a solution and create the file column called `Pdf` in the `Invoice` table with the following steps:

1. Open the `Buildapp` solution.

2. Open the `Invoice` table.

3. In the `Invoice` table, create a new column of the file type and call it `Pdf`.

> **Tip**
>
> Further reading on using file column data: `https://learn.microsoft.com/en-us/power-apps/developer/data-platform/file-column-data?tabs=sdk`

Word template

Sarah is collaborating closely with Adam from the accounts department of her client to streamline the process of creating PDF files. Adam shares a Word template they had previously used for mail merging invoices. Together, they review the template, examining each field and section. Notably, the template includes a repeating section dedicated to invoice details. This section serves as a placeholder for a table, structured with columns corresponding to the various invoice details records. Adam confirms that he has meticulously ensured alignment between the fields in the Word template and the corresponding fields in Dataverse for both the invoice record and the invoice details records.

Adam uploads the Word template to a designated OneDrive storage location and grants Sarah access to the folder. Equipped with access to the template, Sarah embarks on the next phase of her project: developing the Power Automate cloud flow responsible for generating the PDF file. With the template readily available, Sarah can seamlessly integrate it into her automation workflow, streamlining the process of creating PDF invoices.

> **Tip**
>
> Further reading on how to use a Word template: `https://learn.microsoft.com/en-us/power-platform/admin/using-word-templates-dynamics-365`

Power Automate cloud flow

The objective of this cloud flow is to generate a PDF file and store it within the Dataverse invoice record. Sarah initiates the creation of a Power Automate flow, leveraging the automated cloud flow triggered by Dataverse's **When a row is added, modified, or deleted** trigger. She names it `Upsert InvoicePDF by Word Template`. Adam emphasizes that access to the PDF should be granted only after the invoice has been populated with its details, which would result in the invoice having values such as invoice total value. He outlines that the **status reason** of the invoice would transition from **Draft** to **Filled** once populated with invoice details. Adam clarifies that various methods exist to populate the invoice with invoice details records, including importing Excel information and manually creating records. Regardless of the method used, they ensure that the invoice's status reason would indicate **Filled** when ready. After considering Adam's insights, Sarah formulated a plan for the cloud flow, outlined as follows:

1. **Trigger the invoice cloud flow**: Initiate the cloud flow when the invoice status reason is modified to **Filled**.

2. **List invoice details rows**: Retrieve all relevant rows from the invoice details table.

3. **Initialize an array**: Set up an array named `InvoiceDetailsArray` within the cloud flow.

4. **Loop through records**: Iterate through each invoice details record using an **apply to each** step.

5. **Populate array**: Populate `InvoiceDetailsArray` with data columns from each invoice details record, aligning them with the corresponding columns in the Word template's repeating section for invoice details.

6. **Fill Word template**: Populate the Word template with fields from the invoice, along with related tables' information from the client and the invoice details block.

7. **Convert to PDF**: Utilize the OneDrive Power Automate connector to convert the populated template into PDF format.

8. **Upload PDF**: Upload the generated PDF file to the corresponding invoice record.

9. **Clean up temporary files**: Delete and perform cleanup operations on any temporary files created during the PDF conversion process in OneDrive.

Fill Word template

To fill the Word template, Sarah uses the OneDrive connector in Power Automate. Here's how she does it:

- **Populate Microsoft Word template**: Sarah adds the **Populate a Microsoft Word template** action. In this step, she selects the Word template stored in OneDrive. She then maps the fields from the Dataverse records to the corresponding placeholders in the Word template.

- **Invoice details**: Sarah uses the invoice and client data retrieved earlier to fill in the placeholders for fields such as invoice number, client name, address, invoice date, due date, and total amount.

- **Repeating section for invoice items**: She maps `InvoiceDetailsArray` to the repeating section in the Word template designated for itemized invoice details, ensuring each item is correctly formatted and placed in the document.

Convert to PDF

After populating the Word template, the next step is to convert this document into a PDF:

- **Create Word file**: Using the **Create file** action, Sarah saves the populated Word template as a new Word document in OneDrive.

- **Convert file**: Sarah then adds the **Convert file** action from the OneDrive connector to convert the newly created Word document into a PDF. She selects the format as PDF and specifies the path of the Word document created in the previous step.

This conversion process ensures that the final output is a PDF file that retains all the formatting and data integrity from the original Word template.

Having completed her cloud flow design, Sarah begins to implement the cloud flow, with her first step being the trigger step, as shown in *Figure 13.3*. Sarah configures this step to run when an invoice is modified. By adding **Filter Rows**, which ensures that the flow would only run under the `statuscode eq 100000004` condition, Sarah opens the default solution in her development environment to retrieve and copy the status reason values of the `Invoice` table for **Filled status reason**, which has a value of `100000004`:

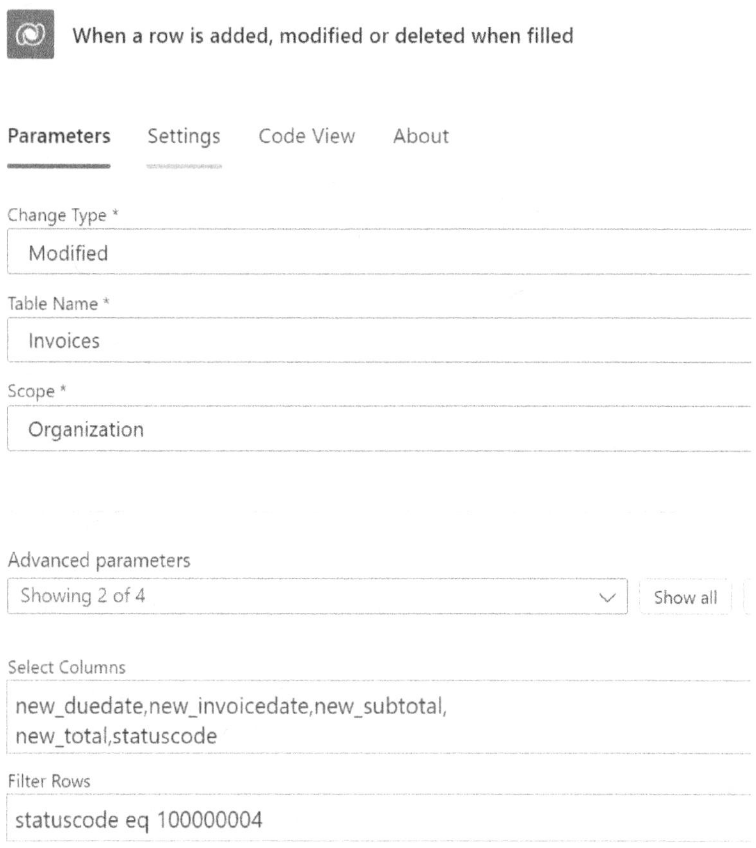

When a row is added, modified or deleted when filled

Parameters   Settings   Code View   About

Change Type *

Modified

Table Name *

Invoices

Scope *

Organization

Advanced parameters

Showing 2 of 4   ∨ | Show all |

Select Columns

new_duedate,new_invoicedate,new_subtotal,
new_total,statuscode

Filter Rows

statuscode eq 100000004

Figure 13.3 – Dataverse trigger step when invoice is filled

Initially, Sarah configures the trigger to activate only when the status code changes. However, upon Adam's advice regarding potential changes to the invoice or its details records, including updates to its value or terms, she expands the trigger's selection criteria to include additional fields such as `new_due`, `new_invoice`, and `new_subtotal`. Additionally, Sarah incorporates **Get Row by ID** steps to retrieve both the invoice and client records, as shown in *Figure 13.4*. While she recognizes that obtaining the invoice record might not be strictly necessary, having it within the flow facilitates testing and direct extraction of information when needed.

To ensure that all relevant client information is available for the invoice, Sarah added a step to get the client record from the `Accounts` table, as shown in *Figure 13.4*. This step is crucial for retrieving the client's address and other related data to be included in the invoice Word template.

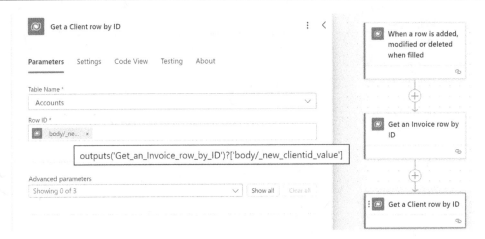

Figure 13.4 – Get a Client row by ID

The client row is an **Accounts** table row, and Sarah selects the row ID from the `Invoice` table. She will use the client record for information such as the client's address and other data that is placed on the invoice Word template. Now she is ready to develop the array that will hold the invoice details in a format that could be placed on the invoice Word template.

Sarah adds a list rows step that she calls List rows `Invoice Details`, as shown in *Figure 13.6*, and she uses the **Fetch Xml Query** field to fill in the query. To get the fetch XML code, Sarah carries out the following steps:

1. Open the advanced find tool in Dataverse, as shown in *Figure 13.5*.

2. Look for the invoice details table, which displays **Invoice Equals** and then the first invoice that comes up as a placeholder in the query.

3. In the configuration of **Edit Columns**, select all the columns needed for the invoice details array.

4. Using the **Download Fetch XML** button, save the fetchxml file to the `Downloads` folder.

5. Open up the file, then copy the fetchxml code and paste it into the **Fetch XML Query** field, as shown in *Figure 13.6*.

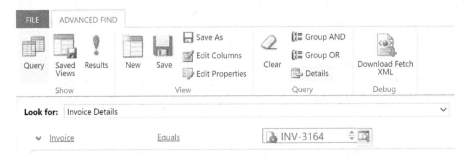

Figure 13.5 – Use advanced find to download fetch XML

Remember to replace the placeholder invoice name and value in the filter condition of the fetchxml and select the value from the invoice ID from the trigger step, as can be seen in *Figure 13.6*. The code looks like this:

```
<condition attribute="new_invoiceid" operator="eq" uitype="new_
invoice" value=" triggerOutputs()?['body/new_invoiceid']" />
```

Sarah configures the **List rows INVOICE DETAILS** step to retrieve the necessary data using the Fetch XML query, as shown in *Figure 13.6*.

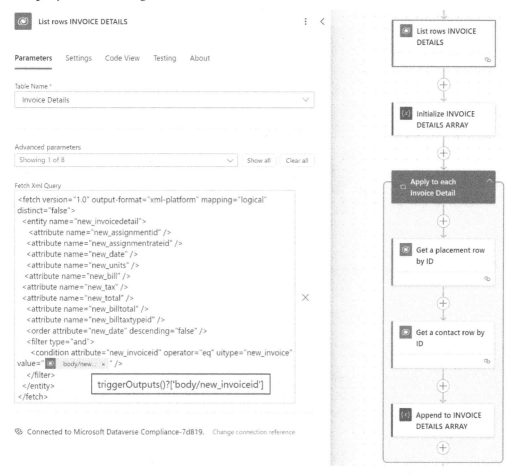

Figure 13.6 – Apply to each Invoice Detail

Sarah wants to have related information also populated in the invoice details array. To do this, she adds some columns related to the invoice details record in the loop, as seen in *Figure 13.6*, including **Get a contact row by ID**. In the **Append to INVOICE DETAILS ARRAY** step, Sarah inputs a JSON block to represent one row of invoice details that would be placed in the PDF template, as shown in *Figure 13.7*.

Figure 13.7 – The invoice details array JSON

The JSON code in *Figure 13.7* represents a template for appending individual invoice details records to an array named `InvoiceDetailsArray`. This array will later be used to populate a Word template for generating a PDF. Here is a detailed explanation of each part of the JSON code:

- **Candidate Name:** This field retrieves the full name from the associated contact record. It uses a dynamic content placeholder to get the value.

- **Rate:** This field pulls the rate associated with the invoice details record.

- **Date:** This field formats the date from the invoice details record. It uses an `if` function to conditionally format the date.

- **Unit:** This field retrieves the number of units from the invoice details record.

- **Unit Price:** This field retrieves the unit price from the invoice details record.

- **VAT**: This field retrieves the VAT amount from the invoice details record.

- **Sub-Total**: This field retrieves the subtotal amount from the invoice details record.

- **Tax**: This field retrieves the tax amount from the invoice details record.

- **Total Amount**: This field retrieves the total amount from the invoice details record.

For each field, Sarah selects the field to match the corresponding JSON label so that as the flow loops through every invoice detail, it will populate and append the array with another item representing an invoice details record to display on the PDF.

Now that Sarah has all the data ready, she can fill the Word template with the invoice information, as shown in *Figure 13.8*.

Sarah browses through the OneDrive storage folder, selects the invoice Word template, and matches up the columns of data with fields on the Word template, including setting the invoice details array block to the Word template, repeating the block labeled **447747060 Item**.

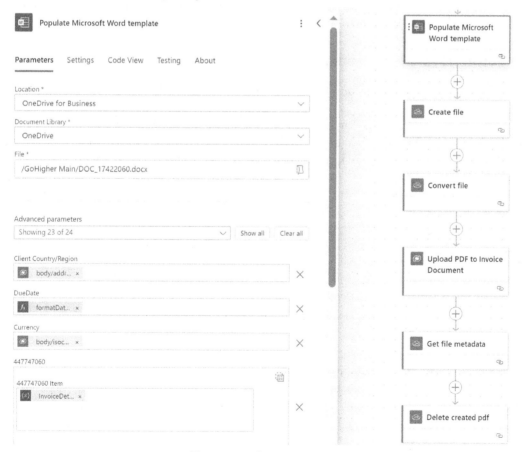

Figure 13.8 – OneDrive steps

To complete the flow to create the PDF, as shown in *Figure 13.8*, do the following:

1. Select a OneDrive step to create a Word document from the Word template content.
2. Select a OneDrive step to convert the Word document into a PDF.
3. Use a Dataverse upload image step to upload a PDF file to the invoice record's PDF column.
4. Use a OneDrive step to get the file metadata of the Word document created in *step 2*.
5. Use the file metadata to delete the Word document as it is no longer needed.

> **Tip**
>
> Further reading on the upload file step in a Power Automate flow: `https://learn.microsoft.com/en-us/power-automate/dataverse/upload-download-file`

Sarah will test the flow by triggering an existing invoice by changing its status to **Filled**. Once that is tested and complete, Sarah needs to place the PDF file on a web page so that authorized users can access it. She will do this as described in the next section by injecting HTML onto a basic form web page.

Injecting HTML into a basic form web page

Sarah will need an object on the form to inject the HTML containing the PDF link so that it can be identified and have the HTML appended at that point on the Dataverse form. Sarah decides to use a section, as shown in *Figure 13.9*. Sarah implements this in the solution; she opens the invoice form and adds a new section with only its label, as shown in *Figure 13.9*. For the label, Sarah inputs `<p id="downloadpdf"></p>`, which will then be appended with the new HTML as can be seen in the JavaScript.

Figure 13.9 – Create a section to append HTML

With the HTML section created, Sarah has set up a designated space on the form for the PDF download link, ensuring a precise location for dynamic content injection. This setup allows her to proceed with integrating the PDF download functionality using Liquid and JavaScript. This next step will ensure that users can easily access and download the generated PDF invoices from the web page.

Liquid

This is the Liquid code placed at the top of the web template. Its main objective is to retrieve the PDF filename, PDFfileName, associated with a specific invoice from Dataverse and then dynamically generate a hyperlink for downloading the PDF file:

```
{% if request.params['invoiceid']%}
{% assign invoiceId = request.params['invoiceid'] %}
{% assign invoice =  entities['new_invoice'][invoiceId] %}
{% assign name = invoice.new_name %}
{% assign PDFfileName  = invoice.new_pdf.Name %}
```

The Liquid block retrieves the invoice ID from the request parameters and then fetches the corresponding invoice details from Dataverse. Additionally, it extracts the name and PDF filename associated with the invoice. The PDF filename (PDFfileName) is then assigned to a variable for later use in the JavaScript block.

JavaScript

This code retrieves the PDF filename associated with a specific invoice and utilizes it to dynamically generate a downloadable PDF link within the designated section of the web page:

```
var entityID = "{{ invoiceId }}";
  var PDFfileName = "{{ pdfName }}";
  var downloadSection = document.getElementById('downloadpdf');
  // Create and append the hyperlink if the element is found
  if (downloadSection && entityID && entityID.trim() !== "" &&
PDFfileName && PDFfileName.trim() !== "") {
    var organizationurl = "{{ settings['OrgUrl'] }}";
      var entitySetName = "new_invoices";
      var fileTypeAttributeName = "new_pdf";
      var pdfUrl = `${organizationurl}/
api/data/v9.1/${entitySetName}
(${entityID})/${fileTypeAttributeName}/$value?size=full`;
        if (downloadSection) {
        // Create the <h3> element
        var h3 = document.createElement("h3");
        h3.textContent = "Download PDF";
         var link = document.createElement("a");
            link.href = pdfUrl;
            link.textContent =  "{{ PDFfileName }}";
            link.target = "_blank";
            downloadSection.appendChild(h3);
            downloadSection.appendChild(link);
        } else {
```

```
            console.log("Download section not found, link not
  appended");
        } }
```

The JavaScript block initializes the `PDFfileName` variable using the PDF filename obtained from the Liquid block.

It constructs the URL for the PDF file using the organization URL, entity set name, and entity ID.

The organization URL is retrieved from a site setting with `{{settings['OrgUrl']}}`; this is good practice as there is no way to retrieve the organization URL through the environment or page otherwise.

The code then dynamically creates an HTML anchor tag, `<a>`, for the PDF download link, setting its `href` attribute to the constructed PDF URL.

The PDF filename is used as the text content of the anchor tag, making it visible to the user.

Finally, it appends the anchor tag to the designated section on the form identified by the ID `downloadpdf`.

This is what it looks rendered on the page:

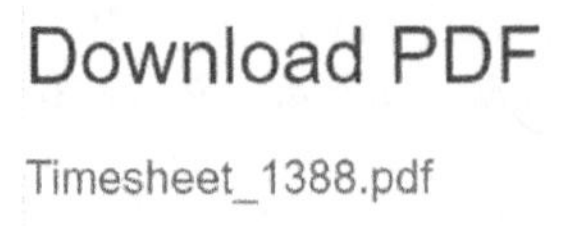

Download PDF

Timesheet_1388.pdf

Figure 13.10 – Rendered download link

The PDF is now created when the status is **Filled**, and the client is satisfied.

Summary

Sarah embarked on a journey to create PDF files from Dataverse invoices following a client request. She focused on enabling customers to download PDF versions of invoices from a web page. Throughout the process, she considered various design choices, such as utilizing Dataverse file fields, leveraging an Office Word template, integrating it with a Power Automate cloud flow, integrating OneDrive, and injecting the PDF link into a web page.

By leveraging tools and technologies effectively, Sarah successfully fulfilled the client's requirements and enhanced the overall workflow efficiency.

In the next chapter, Sarah will build a modal window and connect to a service for checking addresses.

14
Modal Windows

In this chapter, Sarah explores the concept of modal windows within Power Pages. A **modal window** is a type of **user interface** (**UI**) element that temporarily interrupts the workflow of the main application to display critical information or interact with the user.

In the context of Power Pages, modal windows stand as a pivotal component in enhancing **user experience** (**UX**) and interface design. These windows, which appear over the main content of a page, offer a unique opportunity for Sarah and other developers to create interactive and engaging web applications. By integrating modal windows, Sarah can display forms, information, or confirmations without detracting from the primary content, maintaining a seamless and unobtrusive UX. This chapter delves deeper into the practical applications of modal windows in Power Pages, highlighting their versatility and effectiveness in providing contextual interactions, which are essential for creating intuitive and user-friendly web applications.

Specifically, in Power Pages, modal windows are commonly used to present forms or additional content in a focused manner without navigating away from the current page. For example, Sarah has already utilized modal windows as basic forms opened from a button on a page, which triggers the modal window to appear, allowing users to input data or perform specific actions without leaving the context of the page. Sarah had opened Timesheet Costs forms as a modal window from a Timesheet page.

Sarah is going to further explore how she can work with modal windows to deliver improved UX.

In this chapter, Sarah is going to do the following:

- Access variables from the parent page to achieve filtered lookups on a modal window basic form
- Develop a custom modal window to display external data to populate the parent page

Sarah had encountered this problem on most of her basic forms that were rendered as modal windows, where she had data in the parent page but was not able to use it in the basic form. Finally, Sarah was able to solve it, as shown in the next section.

Basic form filtered lookup

Sarah prepared an Agile story to share with her client and to formalize the use case requirement.

Agile user story 1 – filtering placement rates in Timesheet Costs form

Title: Filter Placement Rates in Timesheet Costs Form

As a: Web Developer

I want: to implement filtered lookups in the Timesheet Costs form

So that: users can only see their specific placement rates and not all available rates, ensuring data confidentiality and simplifying the selection process.

Acceptance criteria

Here are the acceptance criteria:

1. **Filtered lookup configuration**:

   - Configure the `Placement Rate` lookup column to be filtered based on the selected `Placement` in the Timesheet Costs form.

   - Ensure the filtered lookup only shows rates related to the selected `Placement`, reflecting the worker's contract rates.

2. **User selection flow**:

   - Automatically populate the `Placement` lookup field when the Timesheet Costs form is loaded.

   - Ensure the `Placement Rate` lookup is filtered based on the pre-populated `Placement` field.

3. **Usability and testing**:

   - Test the Timesheet Costs form to ensure the `Placement` lookup field is pre-populated and read only.

   - Verify that the `Placement Rate` lookup is correctly filtered and user friendly, avoiding the need for users to manually select their `Placement` record.

Design Filtered Lookup on basic form modal window

In the Timesheet Pages, there is a button called new **Timesheet Cost**, which opens a modal window form to add a new Timesheet cost; these are timesheet details containing the date and number of hours worked on that day, as shown in *Figure 14.1*.

Date	Rate Type	Units	Bill Unit	Pay Rate	Bill Rate	Pay Total	Bill Total	Bill TAX Type	
25/02/2024	Standard Day	1	Days	$10.00	$20.00	$10.00	$20.00	NO VAT (0%)	⌄

Figure 14.1 – The Timesheet page with subgrid and basic form button

The button, placed above the subgrid of Timesheet Costs, opens a basic form for the timesheet Costs form and allows the user to enter a specific record for Timesheet Costs then when submitted creates a record on the subgrid, as shown in *Figure 14.2*:

Figure 14.2 – Basic form for Timesheet Costs

The problem is that the Timesheet Costs basic form had two lookups on it: It had the Placement lookup of the timesheet and a Placement Rate lookup, as shown in *Figure 14.2*. Placement Rate had differing rates for overtime, weekend rate, and daily rate. The placement rates are specific to that worker as specified by the Placement record. Sarah could not show all the placement rates; firstly, this is confidential information between the employer and the worker, and also, it would be too difficult for the worker to search for their own placement rate. It was expected that the user would select their placement rate, and the date of the timesheet Costs, and enter the number of units, specifically the number of hours worked or the number of days worked.

To solve the filtering problem of placement rates, Power Pages, through Dataverse forms, allows for **filtered lookups**, where one lookup can filter another, so, therefore, placement rates could be filtered to the placement of the timesheet; in commercial terms, that is the placement rate is filtered to the worker's contract, showing their contracted rates only. Sarah achieved this when she created the Timesheet Costs Dataverse form in the Dataverse form designer. Sarah had placed both the Placement lookup column and the Placement Rate lookup column on the Dataverse form and then configured the Placement Rate lookup column to filter by related rows to the Placement lookup column by selecting **Filter by related rows**, as shown in *Figure 14.3*.

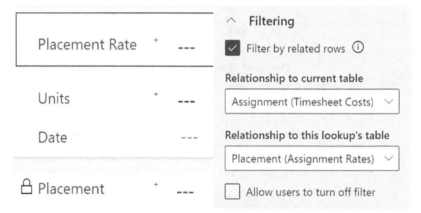

Figure 14.3 – Configuring the Placement Rate lookup column filtering

This solution worked fine; if the user selected Placement, then the Placement Rate values offered in the lookup were now filtered by the worker's placement (e.g., by their contracted rates). Two problems emerged from this: it now required the user to select their own placement record, which was difficult when faced with hundreds of Placement records, and also that the Placement record had to be selected first before Placement Rate was selected.

The timesheet record, which is the parent page containing the timesheet record, contains all the relevant data of the Placement and the worker's candidate record. The solution to this problem and to enable a smooth and easy UX would be to automate the Placement lookup so that it was automatically populated by the parent timesheet record, as the Timesheet Costs form was loaded and then this would allow the Placement record to be read only.

Sarah solved this with the following pattern:

1. She added Liquid code on the timesheet's web template to set the `Placement` record's ID and name.

2. She set the `Placement` record's ID and name as global variables in JavaScript.

3. In the Timesheet Costs basic form, she used JavaScript to populate the `Placement` lookup column.

4. In the Timesheet Costs basic form, she used JavaScript to set the `Placement` lookup column as read only.

With the user story completed and design choices made, Sarah began her implementation.

Implementing the filtered lookup on a basic form modal window

In the following subsections, we will cover the specific steps Sarah took to automate the `Placement` lookup field and solve the filtering problem for placement rates. Here's what each subsection will cover:

1. Liquid code in the Timesheet's web template:

   - Sarah used Liquid code to set the `Placement` record's ID and name dynamically based on the `Timesheet` record.

   - This code ensures that the necessary data is available when the Timesheet Costs form is loaded.

2. JavaScript code in the Timesheet web template:

   - Sarah assigned the `Placement` record's ID and name as global JavaScript variables.

   - This setup allows these values to be accessed and used in other parts of the form, specifically within the modal window.

3. JavaScript code in the basic Timesheet Costs form:

   - Sarah used JavaScript in the Timesheet Costs form to automatically populate the `Placement` lookup field with the values obtained from the parent `Timesheet` record.

   - She also set the `Placement` lookup field to read only to prevent user changes, ensuring data integrity and reducing user effort.

Liquid code in the Timesheet's web template

Sarah had previously created a web template for the Timesheet to support all the code she needed for the pages and timesheet forms. Sarah usually implemented this pattern: whenever she built a page, she would create a web template with the same name as the page and place all her Liquid and JavaScript code into that web template.

Sarah developed the following Liquid code:

```
{% if request.params['timesheetid']%}
{% assign timeSheetId = request.params['timesheetid'] %}
{% assign timeSheet =  entities['new_timesheet'][timeSheetId] %}
{% assign new_assignmentid = timesheet.new_assignmentid.id  %}
{% assign new_assignmentname = timesheet.new_assignmentid.name  %}
```

The preceding code has the following lines:

- `{% if request.params['timesheetid']%}`: This line checks whether there is a parameter named `'timesheetid'` in the URL query string. It ensures that the subsequent code is executed only if this parameter is present.

- `{% assign timeSheetId = request.params['timesheetid'] %}`: Here, the value of the `'timesheetid'` parameter from the URL query string is assigned to the `timeSheetId` variable. This variable will store the unique identifier of the timesheet record.

- `{% assign timeSheet = entities['new_timesheet'][timeSheetId] %}`: This line retrieves the timesheet record from the Dataverse entities using its unique identifier (`timeSheetId`). The timesheet record is stored in the `timeSheet` variable, allowing access to its attributes and related data.

- `{% assign new_assignmentid = timesheet.new_assignmentid.id %}`: This code extracts the unique identifier of the placement associated with the timesheet. It accesses the `'new_assignmentid'` attribute of the timesheet record and assigns its value to the `new_assignmentid` variable.

- `{% assign new_assignmentname = timesheet.new_assignmentid.name %}`: Similarly, this line retrieves the name of the placement linked to the timesheet. It accesses the `'name'` attribute of the `'new_assignmentid'` attribute within the timesheet record and assigns its value to the `new_assignmentname` variable.

Overall, this Liquid code snippet dynamically fetches details of a timesheet, such as its unique identifier, associated placement identifier, and placement name, allowing for customization and automation based on this information within the Power Pages environment.

Sarah now had the `Placement ID` and `Placement` record name assigned to Liquid variables she would need to set them as global JavaScript variables. Sarah would do this in the JavaScript code in the Timesheet web template.

JavaScript code in the Timesheet web template

Sarah created a script section in the web template to contain all the JavaScript used on the Timesheet page and Timesheet form. Sarah created `$(document).ready(function()`, as shown next:

```
<script type="text/javascript">
    $(document).ready(function() {
  // Assign Liquid variables to global JavaScript variables
    window.parentNewAssignmentId = "{{ new_assignmentid }}";
    window.parentNewAssignmentName = "{{ new_assignmentname }}";
    });
</script>
```

The preceding code has these lines:

- `$(document).ready(function()  {`: This line signifies the beginning of a jQuery function that executes when the document (i.e., the web page) is fully loaded and ready to be manipulated.

- `window.parentNewAssignmentId = "{{ new_assignmentid }}";`: Here, the value of the `new_assignmentid` Liquid variable is assigned to a global JavaScript variable named `parentNewAssignmentId`. The Liquid variable is enclosed within double curly braces (`{{ }}`), indicating that its value will be dynamically inserted into the JavaScript code when the page is rendered.

- `window.parentNewAssignmentName = "{{ new_assignmentname }}";`: Similarly, this line assigns the value of the `new_assignmentname` Liquid variable to a global JavaScript variable named `parentNewAssignmentName`. Like before, the Liquid variable's value will be substituted into the JavaScript code during page rendering.

- `});`: This line marks the end of the `$(document).ready()` function, closing the function definition.

Overall, this JavaScript code segment ensures that the values retrieved from the Liquid variables in the Power Pages environment are accessible as global variables in the JavaScript scope. These global variables can then be used across different parts of the web page or within other JavaScript functions to perform various operations, such as dynamically populating form fields or interacting with external systems and used within a modal window, specifically used in the Timesheet costs basic form, which Sarah implements in the next section.

JavaScript code in the Timesheet Costs JavaScript section

Every basic form has a JavaScript field where developers can place JavaScript specifically to that basic form. Mostly, Sarah followed the pattern where she placed most of her JavaScript for forms that are implemented as web pages in a web template. This made it easier as she always knew where most of her code was and it made it easier to find, as her Liquid, custom HTML, and JavaScript are all in one place for a form and web page. However, in this case where the form would be run and rendered as a modal window and on its own web page, then Sarah placed the JavaScript in its own basic form JavaScript field.

As with accessing all her code, Sarah would also access the basic form JavaScript by opening the Visual Studio Code editor through the Power Pages studio and browsing through the editor to find the basic forms where all the JavaScript code for each form was located, as shown in *Figure 14.4*:

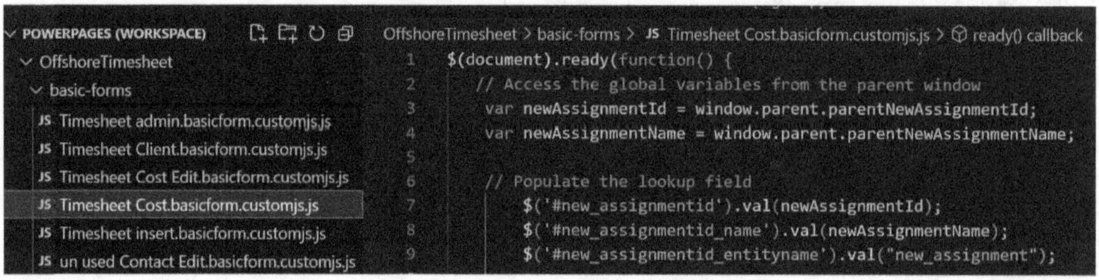

Figure 14.4 - Visual Studio Code basic form JavaScript

Sarah would browse through the Visual Studio Code editor and locate the Timesheet Costs basic form that she had implemented as a modal window that would open when the button to create a new timesheet Costs record was pressed:

```
$(document).ready(function() {
    // Access the global variables from the parent window
    var newAssignmentId = window.parent.parentNewAssignmentId;
    var newAssignmentName = window.parent.parentNewAssignmentName;
    // Populate the Placement lookup field
        $('#new_assignmentid').val(newAssignmentId);
        $('#new_assignmentid_name').val(newAssignmentName);
        $('#new_assignmentid_entityname').val("new_assignment");
});
```

The provided JavaScript code is responsible for populating the `Placement` lookup field in the Timesheet Costs form with values obtained from global variables set in the parent Timesheet web page:

- `$(document).ready(function() {`: This line begins a jQuery function that executes when the document (i.e., the basic form) is fully loaded and ready for manipulation.

- `var newAssignmentId = window.parent.parentNewAssignmentId;`: Here, the `newAssignmentId` JavaScript variable is declared and assigned the value of the `parentNewAssignmentId` global variable from the parent window. This global variable was previously set in the parent Timesheet web page using Liquid code.

- `var newAssignmentName = window.parent.parentNewAssignmentName;`: Similarly, this line declares the `newAssignmentName` JavaScript variable and assigns it the value of the `parentNewAssignmentName` global variable from the parent window. This variable holds the name of the placement associated with the timesheet.

- `$('#new_assignmentid').val(newAssignmentId);`: This line sets the value of the `Placement` lookup field with the `new_assignmentid` ID to the value stored in the `newAssignmentId` variable. It effectively populates the `Placement` lookup field with the ID of the placement obtained from the parent Timesheet web page.

- `$('#new_assignmentid_name').val(newAssignmentName);`: Similarly, this line sets the value of the name of the placement (`new_assignmentid_name`), to the value stored in the `newAssignmentName` variable. To set a lookup, the exact Lookup's record name must be set.

- `$('#new_assignmentid_entityname').val("new_assignment");`: This line sets the value of the table name, the entity name, for the placement lookup. The table name is a required property when setting a lookup field.

Finally, Sarah tested opening the Timesheet Costs basic form as a modal window and saw that the `Placement` lookup was populated on the form load and that the `Placement Rate` lookup was filtered correctly, as shown in *Figure 14.2*.

Sarah was able to use this pattern to populate many columns on different basic forms that were opened as a child modal window from a parent web page and it worked well, delivering a smooth UX and reducing the number of inputs a user needed to enter.

Sarah now had a more complex requirement. Her client had given her an API service that they had subscribed to, which listed corporation addresses and their corporation registration number. By inputting a partial corporation name, for example, entering `Microsoft` and a country, the service would locate all the Microsoft addresses and corporation registration numbers associated with Microsoft in that country. In the next section, Sarah will design and implement a solution to do this with a custom modal window.

A custom modal window for External Data API

Sarah was provided with access credentials and an API designed to retrieve corporation information based on a submitted name or partial name. This API service facilitated the retrieval of data related to addresses and corporate details that matched the provided search criteria. The primary requirement was for users to interact with the returned data in the form of a list, enabling them to scroll through and select a specific row. Upon selecting a row, the data associated with that selection needed to be seamlessly integrated into Sarah's web page, which housed a record for the corporation, including pertinent details such as its address and corporate information.

Such a pattern was not uncommon in commercial projects. For instance, Sarah had a similar project on the horizon that involved utilizing an address finder API service. In this upcoming project, users would utilize an information service to obtain a list of addresses based on a ZIP or postal code entry.

Within Sarah's parent page, several fields required automatic population upon user selection of a row. These fields included `Registration Number` (e.g., Corporation or Company Registration number), `Account Name`, `Street Address`, `Address Locality` (e.g., town), `Address Postcode` (e.g., ZIP), and `Address Country`. It was imperative that the custom modal window featured a scroll bar and displayed a spinner upon loading to enhance UX and provide visual feedback.

As usual, Sarah prepared an Agile user story to share with her client and agree on the use case requirement.

Agile user story 2 – a custom modal window for External Data API

Title: Implement Custom Modal Window for External Data API

As a: Web Developer

I want: to develop a custom modal window that displays external data from an API and populates the parent page form based on user selection

So that: users can easily retrieve and select corporation addresses and registration numbers without leaving the current page.

Acceptance criteria

Here are the acceptance criteria:

1. **Modal window design**:

    *   Create a custom modal window that can be triggered by an event (e.g., a button click) on the parent page.
    *   Ensure the modal window displays a list of corporation addresses and registration numbers based on API data.

2. **API integration**:

    *   Integrate the provided external API to fetch data based on user input (e.g., partial corporation name and country).
    *   Display the retrieved data in the modal window with the appropriate loading indicators and error handling.

3. **User interaction**:

    *   Allow users to scroll through the list of retrieved data and select a specific row.
    *   Populate the parent page form fields (e.g., `Registration Number`, `Account Name`, `Address`) with data from the selected row.
    *   Ensure the modal window can be closed easily by the user and is responsive to different screen sizes.

4. **Testing and validation**:

    *   Test the modal window to ensure it displays data correctly based on API responses.
    *   Validate that the selected data is accurately transferred to the parent page form fields.
    *   Ensure the overall UX is smooth, with clear instructions and feedback mechanisms.

Implementing a custom modal window for External Data API

To implement this solution, Sarah applied a familiar pattern she had mastered while working with cloud flows in Power Pages. She centralized all the API logic and complex operations within a codeless Power Automate cloud flow. This cloud flow was then invoked from the JavaScript code embedded within the web template of the web page. Sarah established a dedicated web template named `AccountDataByWebAPI` to house all the necessary code, including the creation of the custom modal window, populating it with data retrieved from the Power Pages cloud flow response, and managing the selection of rows to populate the parent page's form fields efficiently.

Web template – AccountDataByWebAPI

The provided web template aims to create a custom modal window for retrieving corporation information from an external API. This modal window displays a list of corporations matching the search criteria, allowing users to select one. Upon selection, the modal window populates specific form fields in the parent page with the selected corporation's information. The modal window contains a close button for user interaction and includes a spinner to indicate loading during AJAX requests:

```
<div class="container" role="main">
    <!-- Modal Structure -->
    <div id="Addrs_modalContainer" class="Addrs_modal">
        <div class="Addrs_modal-content">
            <span class="Addrs_close-button">&times;</span>
            <div id="AddrsSVCdetail" style="width: 100%; padding-top:
50px;">
                <div id="loadingSpinner" style="position: absolute;
top: 50%; left: 50%; transform: translate(-50%, -50%);">
                    <!-- Account details will be appended here -->
                </div>
            </div>
        </div>
    </div>
</div>
```

- This HTML structure defines the layout of the modal window.

- It includes a container with the role of `main` for the modal content

- The modal container has an ID of `Addrs_modalContainer` and a class of `Addrs_modal` for styling purposes

- Within the modal, there's a close button (with an **X** symbol), a content area for corporation details, and a loading spinner

The next step involves using JavaScript to control the behavior of this modal window, ensuring it only appears when certain conditions are met and that it seamlessly integrates with the parent page's form fields:

```
$(document).ready(function() {
        $('#new_address1').parent().parent().hide();
        $('#new_addresslocality').parent().parent().hide();
        $('#new_addresspostcode').parent().parent().hide();
        $('#new_registeredcountry, #new_accountname').on('change',
function() {
                console.log("Change event triggered");
                var countryCode = $('#new_registeredcountry_name').val();
                var accountName = $('#new_accountname').val();
                if (countryCode && accountName) {
                    console.log("Change event triggered, passing
parameters");
                    openModalWindow();
                } else {
                    console.log("Conditions not met");  }  });
    });
```

- This JavaScript code is executed when the document is fully loaded and ready for manipulation with `$(document).ready(function()`

- It initially hides certain form fields related to the address

- Event listeners are added to specific form fields (`'new_registeredcountry'`, `'new_accountname'`) to trigger actions when their values change

- If certain conditions are met (e.g., required fields have values), the modal window is opened using the `openModalWindow` function

The openModalWindow() function

The `openModalWindow()` function is designed to manage the display of a modal window on the web page. This modal window serves the purpose of presenting data fetched from an external API. This function ensures that the modal is shown to the user only when necessary and handles the closing of the modal as well:

- **Modal display management**: By encapsulating the logic for displaying the modal in a separate function, the code base becomes more modular and easier to maintain. It provides a clear and organized way to control the visibility of the modal.

- **User interaction handling**: Event listeners are added to the close button and outside clicks to provide multiple options for users to close the modal. This enhances the UX by accommodating different user preferences for interacting with the modal.

Here is the code for `openModalWindow()`:

```
function openModalWindow() {
        // Retrieve the modal element by its ID
        var modal = document.getElementById("Addrs_modalContainer");
        // Check if the modal is already open; if so, prevent opening
it again
        if (modal.style.display === "block") {
            return;              }
```

Here, the function retrieves the modal element from the HTML document using its `Addrs_modalContainer` ID. It then checks whether the modal is already open by inspecting its display style. If the modal is already displayed (i.e., its style is set to `block`), the function returns early to prevent duplicating the modal:

```
        // Display the modal
        modal.style.display = "block";
```

After ensuring that the modal is not already open, the function sets the display style of the modal to `block`, effectively showing it on the web page. This line retrieves the close button element within the modal using its class name, `Addrs_close-button`:

```
// Retrieve the close button element within the modal
var span = document.getElementsByClassName("Addrs_close-button")[0];
```

Since there could be multiple elements with this class, `[0]` is used to select the first matching element. An event listener is added to the close button element:

```
        // Add an event listener to the close button to hide the modal
when clicked
        span.onclick = function() {
            modal.style.display = "none";              }
```

When the close button is clicked, the `span.onclick` function sets the display style of the modal to none, effectively hiding it from view.

Another event listener is added to the `window` object, listening for clicks outside of the modal:

```
// Add an event listener to hide the modal when clicking outside of it
        window.onclick = function(event) {
            if (event.target == modal) {
                modal.style.display = "none";    }      }
```

If the clicked element is the modal itself (i.e., the modal backdrop), the function hides the modal. This allows users to close the modal by clicking anywhere outside of it, enhancing UX and usability.

Finally, the function initiates an AJAX call to fetch data from an external API, via the Power Pages cloud flow:

```
// Initiate AJAX call to fetch data from the external API
initiateAjaxCall();      }
```

This call is made after the modal is displayed to the user, ensuring that the UX is not interrupted by loading data in the background.

The initiateAjaxCall function

The `initiateAjaxCall()` function is designed to handle the initiation of an AJAX call to fetch data from an external API. This function is crucial for retrieving data asynchronously without refreshing the entire web page. It also incorporates error handling and feedback mechanisms to ensure a smooth UX:

- **Feedback with Loading Spinner**: Displaying a loading spinner informs users that data is being fetched, reducing uncertainty and providing feedback about the ongoing process.

- **Error Handling**: In case of a failed request, the function provides feedback to the user with a warning message. Additionally, it ensures that the loading spinner is hidden and the modal window is closed to prevent any UI inconsistencies.

Here is the code for `initiateAjaxCall`:

```
    function initiateAjaxCall() {
// Display a loading spinner while waiting for the response
        var spinner = document.getElementById('loadingSpinner');
        spinner.style.display = 'block';
```

This section of the code retrieves the loading spinner element from the HTML document using its `loadingSpinner` ID and sets its display style to "block". This action makes the loading spinner visible to the user, indicating that data retrieval is in progress.

Here, the `initiateAjaxCall` function retrieves the values entered by the user in the form fields with IDs, `new_registeredcountry` and `new_accountname`:

```
// Retrieve values from form fields for API request
        var CountryCode_Value = $('#new_registeredcountry').val();
        var Name_Value = $('#new_accountname').val();
```

These values are essential parameters for the API request, allowing the function to customize the request based on user input.

This part of the code prepares the data and payload for the AJAX request:

```
var _url = "https://XXX.powerappsportals.com/_api/cloudflow/v1.0/
trigger/ade667ff-68d8-bcc5-c190-2052f6f47f78";
        // Prepare data and payload for the AJAX request
        var data = { "CountryCode_Value": CountryCode_Value, "Name_
Value": Name_Value };
        var payload = { "eventData": JSON.stringify(data) };
```

The data object contains the parameters required for the API request, while the payload object encapsulates the data object in JSON format, as expected by the API endpoint.

Here, the function makes an AJAX POST request to the specified URL (_url) using the ajaxSafePost() jQuery method provided by the shell object:

```
// Perform an AJAX POST request to the specified URL
shell.ajaxSafePost({
contentType: "application/json",
url: _url,
data: JSON.stringify(payload),
processData: false,
global: false,
        })
```

The AJAX POST request includes parameters such as type (POST), contentType (application/json), URL, data (payload in JSON format), processData (false, to prevent jQuery from automatically processing the data), and global (false, to prevent triggering global AJAX event handlers).

Upon successful completion of the AJAX request, the .done() method is executed:

```
.done(function(response) {
        // Hide the loading spinner when the response is received
        spinner.style.display = 'none';

        // Process the received response data
        processAccountsAddress(response);          })
```

The .done() method hides the loading spinner by setting its display style to none and passes the response data to the processAccountsAddress() function for further processing.

If the AJAX request fails, the .fail() method is executed:

```
.fail(function () {
        // Hide the loading spinner and display a warning message
on failure
        spinner.style.display = 'none';
```

```
            alert("Warning: Identity Resolution - No match found for
account " + Name_Value);
```

The `.fail()` method hides the loading spinner, displays a warning message using an alert, and hides the modal window to provide feedback to the user about the failure:

```
            // Hide the modal window
            var modal = document.getElementById("Addrs_
modalContainer");
            modal.style.display = "none";          });       }
```

The processAccountsAddress(response) function

The `processAccountsAddress(response)` function is responsible for processing the response received from the AJAX call, extracting account details, and displaying them in the UI. It involves parsing JSON data, dynamically generating HTML elements, and handling user interactions such as selecting an account:

- **Dynamic data rendering**: The function dynamically generates HTML elements to display account details based on the data received from the server. This approach allows for flexibility and scalability in presenting the information.

- **User interaction handling**: Event listeners are added to the generated HTML elements to handle user interactions, such as selecting an account or canceling the action. This ensures a responsive and intuitive UX.

- **Error handling**: The function includes error handling mechanisms to address scenarios where the response data cannot be parsed or no accounts are found. This helps in providing appropriate feedback to the user in case of errors.

Here is the code for `processAccountsAddress(response)`:

```
            function processAccountsAddress(response) {
  var spinner = document.getElementById('loadingSpinner');
            var AddrsSVCdetaildiv = document.
  getElementById("AddrsSVCdetail");
```

This section retrieves the loading spinner and the main detail div from the HTML document using their respective IDs. These elements are essential for showing and hiding the loading spinner and appending generated account details.

Here, the function checks whether the main detail div (`AddrsSVCdetaildiv`) exists in the DOM:

```
            if (!AddrsSVCdetaildiv) {
  console.error("AddrsSVCdetail div not found in the DOM");
            spinner.style.display = 'none';
            return; }
```

If it doesn't, an error is logged, the loading spinner is hidden, and the function exits to prevent further execution without the necessary element:

```
spinner.style.display = 'none';
```

After ensuring that the main detail div exists, the loading spinner is hidden to indicate that the data processing is complete. This part of the code attempts to parse the JSON response received from the server:

```
console.log("Full Response:", response);
var result;
try {
    result = JSON.parse(response);
} catch (e) {
    console.error("Error parsing JSON:", e);
    return;
}
```

If successful, the parsed result is stored in the `result` variable. Any parsing errors are caught and logged, and the function exits. Next, the function retrieves the `accounts` data from the parsed response:

```
var accountsString = result.matchaccounts;
console.log("Accounts String:", accountsString);
var accounts;
try {
    accounts = JSON.parse(accountsString);
    console.log("Parsed Accounts:", accounts);
} catch (e) {
    console.error("Error parsing matchaccounts:", e);
    return;             }
```

The `matchaccounts` property contains the JSON string representing the accounts. This string is parsed to obtain the accounts array.

The `forEach` function iterates over each account in the `accounts` array. `function(account, index)` is a callback function that executes for each account in the array. It takes two parameters: the current `accounts` object and `index` in the array:

```
                    AddrsSVCdetaildiv.innerHTML = '';
  if (accounts.length > 0) {
      accounts.forEach(function(account, index) {
```

`if (account.organization)` checks whether the current account has an `organization` property:

```
    if (account.organization) {
```

The preceding ensures that only valid accounts with organization details are processed.

```
var org = account.organization;
var div = document.createElement('div');
div.className = 'account-info';
div.id = 'account-' + index;
```

In the preceding code, `var div = document.createElement('div');` creates a new `div` element for displaying account information, `div.className = 'account-info';` assigns the class name `account-info` to the `div` element for styling purposes, and `div.id = 'account-' + index;` assigns a unique ID to the `div` element based on the current index. This helps identify each account div uniquely.

```
// Set data attributes for the account
div.setAttribute('data-registration-number', org.
registrationNumbers && org.registrationNumbers.length > 0 ? org.
registrationNumbers[0].registrationNumber : '');
div.setAttribute('data-primary-name', org.primaryName ||
'');
div.setAttribute('data-primary-address', org.
primaryAddress ? `${org.primaryAddress.streetAddress.line1},
${org.primaryAddress.addressLocality.name}, ${org.primaryAddress.
postalCode}, ${org.primaryAddress.addressCountry.isoAlpha2Code}` :
'');
```

In the preceding code, `div.setAttribute('data-registration-number', ...)` sets a data attribute to store the registration number of the account. It checks whether `org. registrationNumbers` exists and is not empty before setting the attribute.

`div.setAttribute('data-primary-name', ...)` sets a data attribute to store the primary name of the account.

`div.setAttribute('data-primary-address', ...)` sets a data attribute to store the primary address of the account. It constructs the address string using the primary address details if available.

```
div.innerHTML = `
    <div class="Addrs_Row">
        <div class="Addrs_Column Addrs_left-align">
                <strong>account Name</strong>
            </div>
        </div>
    <div class="Addrs_Row">
    <div class="Addrs_Column Addrs_left-align">
      ${org.primaryName || ''}</div>
    <div class="Addrs_Column Addrs_right-align">
                Registration Number: ${org.
```

```
registrationNumbers && org.registrationNumbers.length > 0 ? org.
registrationNumbers[0].registrationNumber : ''}
                            </div>
                        </div>
                        <div class="Addrs_Address">
                            <strong>account Address</strong><br>
                            ${org.primaryAddress ? `${org.
primaryAddress.streetAddress.line1}, ${org.primaryAddress.
addressLocality.name}, ${org.primaryAddress.postalCode}, ${org.
primaryAddress.addressCountry.isoAlpha2Code}` : ''}
                        </div><hr class="Addrs_Divider">`;
```

In the preceding code, `div.innerHTML = ...` sets the inner HTML content of the `div` element to display account details. It includes account name, registration number, and address information formatted in HTML.

```
                    div.addEventListener('click', function() {
                        var currentlySelected = document.
querySelector('.selected');
                        if (currentlySelected) {
                            currentlySelected.classList.
remove('selected');  }
div.classList.add('selected');
                    });
```

In the preceding code, `div.addEventListener('click', ...)` adds a click event listener to the `div` element. When the user clicks on an account div, this function is triggered.

Within the event listener function, it first removes the `selected` class from any previously selected account div and then adds the `selected` class to the clicked account div.

`AddrsSVCdetaildiv.appendChild(div);` appends the created `div` element to the main detail div (`AddrsSVCdetaildiv`).

```
AddrsSVCdetaildiv.appendChild(div);                    });
```

The preceding code adds the account information to the UI for display.

`var buttonsDiv = document.createElement('div');` creates a new `div` element to contain buttons for actions related to account selection:

```
// After appending all accounts
var buttonsDiv = document.createElement('div');
buttonsDiv.className = 'Addrs_Buttons';
```

`buttonsDiv.className = 'Addrs_Buttons';` assigns the `Addrs_Buttons` class name to the buttons container for styling purposes.

`buttonsDiv.innerHTML = ...` sets the inner HTML content of the buttons container. It contains HTML code for the `Cancel` and `Select` buttons:

```
buttonsDiv.innerHTML = `
<button id="cancelButton">Cancel</button>
<button id="selectButton">Select</button>`;
```

In the preceding code, `<button id="cancelButton">Cancel</button>` creates a `Cancel` button.

`<button id="selectButton">Select</button>` creates a `Select` button.

`AddrsSVCdetaildiv.appendChild(buttonsDiv);` appends the buttons container to the main detail div (`AddrsSVCdetaildiv`). This adds buttons to the UI for interaction:

```
AddrsSVCdetaildiv.appendChild(buttonsDiv);
```

`document.getElementById('cancelButton').addEventListener('click', ...)` adds an event listener to the `Cancel` button. When clicked, it hides the modal window (`modal.style.display = "none";`):

```
// Event listeners for buttons                 document.
getElementById('cancelButton').addEventListener('click', function() {
                modal.style.display = "none";
        });                        document.
getElementById('selectButton').addEventListener('click',
function(event) {
```

`document.getElementById('selectButton').addEventListener('click', ...)` adds an event listener to the `Select` button. When clicked, it triggers a function to extract data from the selected account and populate form fields.

`event.preventDefault();` prevents the default form submission behavior when the `Select` button is clicked. This ensures that the page doesn't reload or submit the form prematurely.

`var selectedDiv = document.querySelector('.selected');` retrieves the currently selected account div by querying for an element with the `selected` class:

```
event.preventDefault();
var selectedDiv = document.querySelector('.selected');
if (selectedDiv) {
// Extracting data from the selected record
var registrationNumber = selectedDiv.getAttribute('data-registration-
number');
var new_accountname = selectedDiv.getAttribute('data-primary-name');
// Extracting address fields
var primaryAddress = selectedDiv.getAttribute('data-primary-address').
```

```
split(', ');
var streetAddress = primaryAddress[0];
var addressLocality = primaryAddress[1];
var postalCode = primaryAddress[2];
var addressCountry = primaryAddress[3];
```

If a selected account div exists, the code proceeds to extract relevant data such as registration number, account name, and address fields from the div's data attributes.

`$('#new_registrationnumber').val(registrationNumber);` sets the value of the registration number field in the parent form to the extracted registration number:

```
// Populating the parent form fields
$('#new_registrationnumber').val(registrationNumber);
$('#new_accountname').val(new_accountname);
// Populate other address fields as needed...
$('#new_address1').val(streetAddress);
$('#new_addresslocality').val(addressLocality);
$('#new_addresspostcode').val(postalCode);
$('#new_addresscountry').val(addressCountry);
```

Similar lines populate other address fields such as account name, street address, locality, postal code, and country based on the selected account's data. The following code snippet closes the modal window:

```
                    // Close the modal
                    var modal = document.getElementById("Addrs_
modalContainer");
                    modal.style.display = "none";     } });
AddrsSVCdetaildiv.style.display = "block";
            } else {
                console.error("No accounts found.");        }       }
```

- The code first clears any previous account details displayed in the main detail div (`AddrsSVCdetaildiv`)

- If accounts are found in the response, it iterates over each account and creates a `div` element for displaying account information

- Data attributes are set on each `div` element to store account details for later use

- HTML content is generated dynamically within each `div` element to display the account name, registration number, and address

- Event listeners are added to each `div` element to handle account selection

- Cancel and select buttons are created and appended to the main detail div

- Event listeners are added to these buttons to handle canceling the action or selecting an account

- If no accounts are found, an error message is logged to the console

Summary

Throughout this chapter, Sarah has gained valuable insights into the practical applications and advantages of using modal windows in Power Pages. Her journey has highlighted several key things:

- **Enhanced UX**: Sarah learned that modal windows could significantly enhance the UX by allowing for interactions without navigating away from the current page. This approach helps maintain the user's context and focus, leading to a more intuitive and efficient user journey.

- **Data interaction and display**: The use of modal windows for displaying forms and additional content has proven to be a powerful tool. Sarah successfully implemented modal windows to show forms such as the Timesheet Costs form, where users can input data efficiently.

- **Filtered lookups and data integration**: A significant thing that Sarah learned was how to use modal windows to handle complex data interactions, such as filtered lookups. By linking modal windows with the parent page data, she achieved a more streamlined and relevant data entry process, enhancing data accuracy and user convenience.

- **Customization and flexibility**: Sarah explored how modal windows could be customized and used for various purposes, from simple data entry forms to displaying external data. This flexibility allows for a wide range of applications, catering to specific user needs and scenarios.

- **Improved UI design**: Modal windows contribute to a cleaner and more organized interface. By using these windows, Sarah was able to keep the main page layout uncluttered, only showing additional information or forms when necessary.

The incorporation of modal windows in Power Pages is a testament to the evolution of UX design. For Sarah, the ability to create focused, contextually relevant interactions without overwhelming the user or disrupting the workflow is invaluable. These windows enhance the usability of web applications, making them more intuitive and efficient. In summary, modal windows not only improve the aesthetic and functional aspects of web applications but also play a crucial role in creating user-centric and responsive designs. In the next chapter, Sarah will learn how to use Copilot and ChatGPT to generate designs and code for customizing Power Pages.

15

Enhancing Development with ChatGPT

As a nascent developer, Sarah's journey with Power Pages has been both challenging and rewarding. Starting from the basics of creating simple web pages and working her way up to integrating complex workflows and custom code, she has come a long way. Throughout this journey, she has encountered numerous obstacles, each teaching her valuable lessons and pushing her to explore new tools and techniques.

In this final chapter, Sarah will share how AI-powered development tools, specifically ChatGPT and Copilot, have revolutionized her development process. These tools have become invaluable allies, providing guidance, generating code snippets, and helping her debug issues more efficiently than ever before.

In this chapter, we'll cover the following topics:

- An overview of ChatGPT and Microsoft Copilot
- The benefits of using AI tools in web development
- Getting started with ChatGPT
- Debugging and testing with AI assistance
- Best practices for using AI in debugging and testing
- Maintaining the documentation of prompts and tasks
- Practical tips and best practices
- Copilot in Power Pages Studio
- Copilot and ChatGPT together

An overview of ChatGPT and Microsoft Copilot

In this section, we'll take an overview of ChatGPT and Microsoft Copilot.

ChatGPT

ChatGPT, developed by OpenAI, is an advanced language model that can understand and generate human-like text. It is capable of assisting with a wide range of tasks, from answering questions and providing explanations to generating code snippets and offering debugging advice. Sarah has found ChatGPT incredibly helpful for brainstorming ideas, writing initial drafts of code, and understanding complex concepts.

Copilot

Copilot, a collaboration between GitHub and OpenAI, is an AI-powered code completion tool that integrates directly into the Visual Studio Code editor. By analyzing the context of Sarah's code, Copilot can suggest entire lines or blocks of code, making her development process faster and more efficient. Whether she is writing HTML, CSS, JavaScript, or any other language, Copilot offers relevant and context-aware suggestions that have significantly reduced the time she spends on repetitive coding tasks.

The benefits of using AI tools in web development

Incorporating AI tools such as ChatGPT and Copilot into web development offers numerous advantages that enhance productivity, learning, code quality, and the overall developer experience. In this section, we will look at efficiency and speed, enhanced learning, improved code quality, reduced cognitive load, and real-time collaboration.

Efficiency and speed

One of the most significant benefits Sarah has experienced is a boost in efficiency. With ChatGPT and Copilot, she can quickly generate boilerplate code, troubleshoot issues, and get suggestions for best practices. This has drastically reduced the time it takes to move from concept to implementation.

Enhanced learning

These tools have also been fantastic learning companions for Sarah. When she encounters a new concept or technology, she can ask ChatGPT for explanations and examples, which helps her understand and apply new knowledge more effectively. Copilot, conversely, exposes her to different coding patterns and techniques, broadening her skill set as she uses it.

Improved code quality

By leveraging AI tools, Sarah has noticed an improvement in the quality of her code. ChatGPT helps her think through problems and come up with robust solutions, while Copilot provides code suggestions that are often optimized and follow industry standards. This dual support ensures that her projects are not only functional but also well-structured and maintainable.

Reduced cognitive load

Web development can be mentally exhausting, especially when dealing with complex logic or debugging stubborn issues. ChatGPT and Copilot help alleviate this cognitive load by providing immediate assistance and reducing the mental effort required to solve problems. This allows Sarah to focus more on creative aspects of development and less on repetitive or frustrating tasks.

Real-time collaboration

Another benefit is the real-time collaboration these tools enable. By integrating seamlessly into her workflow, they act as virtual pair programmers, always available to assist with coding, debugging, and optimizing her projects. This constant support has made Sarah more confident in tackling challenging tasks and exploring new features.

As Sarah reflects on her journey, she realizes that embracing AI-powered development tools has been a game-changer. They have not only made her a more efficient and capable developer but also enriched her learning experience and enhanced the quality of her work. In this chapter, she will delve into the practical applications of ChatGPT and Copilot, sharing how they can be harnessed to streamline your development process and take your skills to the next level.

Getting started with ChatGPT

Leveraging ChatGPT can significantly enhance your development workflow, making it essential to understand how to set up and use this powerful tool effectively.

Setting up ChatGPT

To begin leveraging ChatGPT, Sarah first needed to set up an account. This process was straightforward and started by opening a web browser and navigating to the OpenAI website. Sarah signed up for an account and familiarized herself with the platform's features.

Constructing effective prompts

Sarah quickly learned that the effectiveness of ChatGPT depends heavily on how prompts are framed. Constructing clear and specific prompts ensures that the responses are relevant and useful. She also discovered the importance of maintaining a separate document to refine her prompts and keep track of her conversations. This document became a valuable resource for summarizing her interactions and appending specific questions related to her current tasks.

Best practices for prompts

Here are some suggestions for prompts:

- **Be specific**: Clearly define what you need. Vague prompts often result in less useful responses.
- **Provide context**: Give as much context as possible to help ChatGPT understand your requirements.
- **Iterate and refine**: Start with a broad question and then refine based on the responses you get.
- **Summarize and append**: Keep a summary of the ongoing conversation and append new questions to maintain continuity.

Example – generating a Liquid web template

Sarah decided to use ChatGPT to generate a Liquid template for a blog post layout. She constructed her prompt carefully:

- **Prompt**:

  Generate a Liquid template for a blog post layout with a title, author, date, and content section.

- **Response**:

  ```
  <div class="blog-post">
      <h1>{{ post.title }}</h1>
      <p>by {{ post.author }} on {{ post.date }}</p>
      <div class="content">{{ post.content }}</div>
  </div>
  ```

By framing her prompt clearly, Sarah received a precise and usable code snippet. She then tested this template in her development environment and made minor adjustments to fit her specific needs.

Continuing the chat and refining the code

When Sarah needed further adjustments to her code, she appended specific questions to her initial prompt summary. Here's how she approached it:

- **Prompt construction**:

    I.   Sarah pastes in her summary and the existing code, and then types in her new instructions.

    - For example, she might start with *"Working in a Power Pages web template."*

    II.   Then, she includes her current code, *"Code so far below:,"* and pastes it into the existing code.

    III. Finally, she adds her specific instructions, *"Give HTML code to add a section for tags at the bottom of the blog post."*

- **Understanding and implementing responses**:

    - ChatGPT might provide the complete code again. Sarah has learned not to copy and paste the entire code blindly because AI suggestions can sometimes introduce errors or unintended changes.

    - Instead, she focuses on understanding the suggested code, copying only the relevant parts and reviewing them before making slight modifications if necessary. This approach has rapidly improved her coding skills.

- **The ChatGPT response**:

```html
<div class="blog-post">
    <h1>{{ post.title }}</h1>
    <p>by {{ post.author }} on {{ post.date }}</p>
    <div class="content">{{ post.content }}</div>
    <div class="tags">
        <p>Tags: {{ post.tags }}</p>
    </div>
</div>
```

Sarah reviewed the new section of code provided. She was already using a CSS class, called `buildapptags`, so she pasted the code snippet she needed into her web template and adjusted the class name in the new code, as follows:

```html
<div class="buildapptags">
    <p>Tags: {{ post.tags }}</p>
</div>
```

By following this iterative process, Sarah efficiently refined her code and resolved any issues with ChatGPT's assistance.

By setting up an account, constructing effective prompts, and maintaining a structured conversation, Sarah was able to leverage ChatGPT to enhance her productivity and code quality significantly. Then, she will work with Copilot to develop code for a navigation bar in a Power Pages web template.

Debugging and testing with AI Assistance

Leveraging AI tools such as ChatGPT and Copilot can significantly enhance the debugging and testing process, making it easier to identify and fix issues in your code. Here's how Sarah can use these tools effectively.

Using ChatGPT for debugging advice

ChatGPT can provide valuable insights and suggestions to fix bugs. Sarah can describe the problem she's encountering, and ChatGPT can help diagnose the issue and offer potential solutions.

Here are the steps:

1. **Describe the bug**:

    *   Clearly explain the issue, including any error messages or unexpected behavior

    *   Provide relevant code snippets to give context

    *   List the actions and confirmations already achieved

2. **Get suggestions**:

    *   ChatGPT will analyze the description and code and then suggest possible fixes or improvements.

Example – debugging a JavaScript function

Using ChatGPT to debug code can streamline the troubleshooting process, as illustrated in this example.

Prompt

In a Microsoft Power Pages web template using Liquid and jQuery, I encountered an error in the `GetExternalData()`. JavaScript function. The following is the code for the web template that contains the function causing the error. I have confirmed that the web template is running correctly and that the Liquid variables have values. The error message is `'TypeError: Cannot read property 'json' of undefined'`.

Then Sarah adds the sentence and instruction to ChatGPT: `Please debug it.`

[Sarah pastes in the code of the web template]

Response

Based on the error, it seems that you need to check the fetch URL and ensure that it returns the correct data format. Here are a few things to check:

- Verify that the URL is correct and the endpoint is reachable

- Ensure that the server returns the expected JSON format

- Add error handling to catch issues during the fetch operation

Refining code with AI's suggestions

ChatGPT and Copilot can help optimize and improve existing code by suggesting better practices or more efficient methods. Sarah can use the AI's refactoring suggestions to enhance her code quality.

Here are the steps:

1. Identify the code to improve:

    - Select the code snippet or function that needs optimization

    - Describe the specific goal, such as improving performance or readability

2. Review Copilot's suggestions:

    - Copilot will generate code suggestions in real time

    - Sarah can review and integrate these suggestions into her code base

Example – optimizing a JavaScript function

Using ChatGPT to optimize code can lead to more efficient and faster-running applications, as shown in this example:

Prompt

"I want to optimize this JavaScript function to reduce its runtime. Please suggest improvements."

Here is the code:

```
function fetchData() {
    fetch('https://api.example.com/data')
        .then(response => response.json())
        .then(data => {
            // Process data
            console.log(data);
```

```
    })
    .catch(error => console.error('Error:', error));
}
```

Response

To make the function more efficient, consider using `async` and `await` for better readability and potentially improved performance. This approach will also help manage errors more effectively and streamline the data-fetching process.

Here is the optimized code:

```
async function fetchData() {
    try {
        const response = await fetch('https://api.example.com/data');
        const data = await response.json();
        console.log(data);
    } catch (error) {
        console.error('Error:', error);
    }
}
```

Sarah has gotten into the habit of using AI tools such as ChatGPT and Copilot to help debug her code. In the next section, we will see how she also considered the best practices.

Best practices for using AI in debugging and testing

- **Clear and detailed descriptions**: When describing issues to ChatGPT, be as detailed as possible. Include error messages, the context of the problem, and any relevant code.

- **Iterative refinement**: Use ChatGPT and Copilot iteratively. Start with broad questions and narrow down to specific issues as you refine the code.

- **Validation**: Always validate AI-generated code in a development environment before deploying it to production.

- **Documenting prompts and responses**: Maintain a separate document to track the prompts and responses. This helps in understanding the evolution of the solution and can serve as a reference for future debugging.

By using ChatGPT and Copilot, Sarah can streamline her debugging and testing process, making her development workflow more efficient and effective. These AI tools not only help resolve issues quickly but also enhance Sarah's coding skills by providing insights and best practices.

Maintaining the documentation of prompts and tasks

Documentation allows Sarah to keep a detailed record of the evolution of her project. By tracking the progress and changes made throughout the development process, she can understand what approaches worked and what didn't, as well as why certain decisions were made. She can also reuse prompts, which help save time.

Sarah can look back at her documented prompts and see how her Liquid web template developed over time, identifying which specific changes led to successful results.

Refine and reuse prompts

Well-documented prompts can be refined and reused for similar tasks in future projects, saving time and ensuring consistency in the quality and style of work.

If Sarah needs to generate another dynamic form in the future, she can refer to her documented prompts to quickly generate a similar solution, reducing the time spent on trial and error.

Facilitate collaboration

Detailed documentation makes it easier for Sarah to collaborate with other developers by providing a clear history of a project's development. Team members can understand the context and rationale behind each step.

If Sarah works with a team, she can share her prompt history, helping new team members get up to speed quickly and align with a project's goals and standards.

Debugging and troubleshooting

When encountering issues, having a record of previous prompts and responses can help Sarah identify where things might have gone wrong. This can streamline the debugging process and lead to quicker resolutions.

If a new issue arises, Sarah can review past prompts to see whether similar problems were encountered and how they were resolved, saving time and effort.

Continuous improvement

By documenting her learning process and the solutions she has implemented, Sarah can continuously improve her coding skills and problem-solving techniques.

Reflecting on past projects, Sarah can identify patterns in her work that could be optimized, learning from both her successes and mistakes.

Best practices for maintaining prompt history

Maintaining a prompt history document is particularly useful for larger tasks such as a new feature. During development, Sarah would reuse prompts and add a summary to her prompt to keep the focus on the task at hand.

Using a dedicated document

- Create a dedicated document (e.g., a Google Doc, Word document, or Markdown file) to record all prompts and responses

Sarah can use sections or headings to organize her prompts by task or project phase, making it easy to navigate and review past entries.

Include context and comments

- Add context to each prompt, explaining what she aimed to achieve with it.
- Include comments on the effectiveness of the response and any modifications made.
- After each prompt, Sarah can write a brief note about what the response helped her accomplish and any changes she had to make.

This provides a comprehensive understanding of each step.

Timestamp and version control

- Include timestamps and version control for each prompt to track when changes were made.

Summarize the key learnings

- At the end of each session or major task, summarize the key learnings and takeaways.
- Sarah can add a "*Key Learnings*" section after each major task to highlight important insights and improvements, helping her apply these learnings to future projects.

Organize by project and task

- Keep prompts organized by project and specific tasks within those projects.
- Use folders or sub-sections within the document to separate different projects and tasks, making it easier to find relevant information quickly.

Regularly review and update

- Regularly review and update the document to keep it relevant and useful.

- Sarah can set aside time at the end of each week to review her documentation, make any necessary updates, and ensure it remains a valuable resource.

Next, Sarah will consider the best practices in AI-supported development.

Practical tips and best practices

Incorporating AI tools such as ChatGPT and Copilot into the development workflow can significantly enhance productivity and code quality. However, to make the most of these tools, it's good to follow practical tips and best practices. Here's a detailed guide for Sarah and other developers

Iterative development with AI tools

Leveraging AI tools such as ChatGPT and Copilot in an iterative development process can lead to more efficient and accurate coding, enhancing both productivity and code quality:

1. **Start small and build up**: Sarah can start by asking ChatGPT to generate basic HTML structures before moving on to more complex layouts and functionalities.

> Tip
>
> Begin with simple prompts and gradually increase complexity. This helps in understanding how AI interprets instructions and allows for fine-tuning responses.

2. **Break down large tasks**: Instead of asking Copilot to create an entire web page, Sarah can request it to generate individual components, such as the navigation bar, footer, and content sections, separately.

> Tip
>
> Divide complex requirements into smaller, manageable parts. This makes it easier for AI tools to provide accurate and relevant responses.

3. **Use feedback loops**: Sarah can generate initial code snippets, test them in her environment, and then refine the prompts based on what works and what needs improvement.

> Tip
>
> Regularly review and refine the outputs generated by AI tools. Provide feedback to the AI (in the case of tools such as Copilot) to improve future responses.

4.   **Maintain a continuous dialogue**: Sarah can append specific questions or adjustments to her initial prompts to get more tailored responses. For example, if the first version of a navigation bar isn't quite right, she can ask for specific changes rather than starting from scratch.

> **Tip**
>
> Keep the conversation going with AI tools by iteratively refining prompts and incorporating feedback. Treat it as an ongoing collaboration.

Handling limitations and leveraging strengths

Effectively using AI tools involves understanding their limitations while leveraging their strengths to maximize productivity and enhance code quality:

1.   **Understand the limitations**: Sarah should validate the code provided by ChatGPT and Copilot, testing it thoroughly to ensure that it meets her requirements.

> **Tip**
>
> Recognize that AI tools have limitations and may not always provide perfect solutions. Be prepared to make manual adjustments and use AI as a supplement, not a replacement.

2.   **Leverage AI strengths**: Sarah can use Copilot to quickly generate repetitive code structures and ChatGPT to explain complex concepts or provide documentation snippets.

> **Tip**
>
> Utilize AI tools for tasks they excel at, such as generating boilerplate code, suggesting improvements, and providing documentation.

3.   **Provide clear and specific prompts**: When asking for a CSS layout, Sarah should specify the design requirements, such as class names, colors, fonts, and responsive behavior, to get more accurate results.

> **Tip**
>
> The quality of AI-generated responses heavily depends on the clarity and specificity of the prompts. Provide detailed instructions and context.

4.   **Use error handling and testing**: Sarah can ask ChatGPT for advice on adding error handling in her JavaScript functions and use Copilot to generate unit tests for her code.

> **Tip**
>
> Incorporate error handling and robust testing in the code generated by AI tools to ensure reliability and stability.

Ensuring code quality and maintainability

Maintaining high standards of code quality and ensuring maintainability is crucial when integrating AI tools into the development process:

1. **Code reviews and refactoring**: Sarah should set aside time to review the code, clean up any unnecessary parts, and ensure it adheres to best practices before finalizing it.

> **Tip**
>
> Regularly review and refactor the AI-generated code to ensure that it meets coding standards and is maintainable.

2. **Consistent naming conventions**: Sarah can use style guides and linters to ensure that the code generated by AI tools adheres to the project's naming conventions and formatting standards.

> **Tip**
>
> Maintain consistent naming conventions and coding styles across a project to enhance readability and maintainability.

3. **Comprehensive documentation**: Sarah can ask ChatGPT to generate documentation comments for her code and then customize them to fit her project's context.

> **Tip**
>
> Document code thoroughly, including the purpose of each function, class, and major code block. This makes it easier for future developers to understand and maintain the code.

4. **Modular and reusable code**: Sarah can request AI tools to generate code in a modular fashion, using functions and classes that encapsulate specific functionality.

> **Tip**
>
> Write modular code that can be reused across different parts of a project. This reduces duplication and makes maintenance easier.

5. **Testing and quality assurance**: Sarah can use automated testing tools to write unit and integration tests for the AI-generated code, ensuring high quality and reliability.

> **Tip**
>
> Implement thorough testing for all AI-generated code to ensure that it functions as expected and integrates well with the existing code base.

Copilot in Power Pages Studio

Sarah wanted to create pages and forms using Copilot. She wanted to create a support ticket system for a new client.

An agile user story to develop a customer support system

As a customer support manager, I want to develop a customer support system so that customers can easily submit support tickets and we can efficiently manage and track these tickets.

Acceptance criteria

1. **Support ticket form creation**:

   - **Given** the need for a system to capture customer issues,

   - **When** I create a support ticket form,

   - **Then** the form should include fields such as customer name, issue description, and contact details.

2. **Data storage setup**:

   - Given the need to store submitted support tickets,

   - When I configure the data storage,

   - Then there should be a table or database to store the support ticket information

3. **Multi-step form**:

   - Given the need for detailed ticket information,

   - When I create the support ticket form,

   - Then it should be a multi-step form that collects all necessary details across multiple steps

4. **Form integration**:

- Given the support ticket form is created,

- When I integrate the form into the customer support system,

- Then the form should be accessible and functional for customers to submit tickets

5. **Testing and validation**:

- Given the support ticket system is set up,

- When I test the system,

- Then it should correctly submit and store ticket data, and notify support staff of new tickets

Implementation using Copilot

Sarah opened her website in Power Pages Studio and selected the **Page design** tab. She entered the prompt shown in *Figure 15.1*, but Copilot was unable to produce a page.

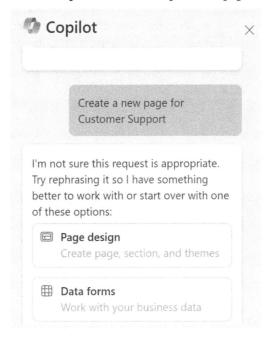

Figure 15.1 – Copilot does not understand what to do

Sarah looked at the examples provided in Copilot and wrote out a similar prompt, as shown in *Figure 15.2*.

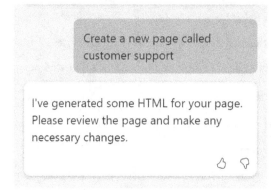

Figure 15.2 – Copilot confirms and produces a web page

Copilot did produce a page, but it created a lengthy page with many sections that resembled an **About us** page design. Sarah could have kept this page and then edited and deleted the unnecessary sections, but she wanted something simple to which she could attach a form to capture user support tickets. Therefore, she decided to specify the number of sections. She tried a simpler prompt, which resulted in the creation of a more straightforward page, as shown in *Figure 15.3*.

Figure 15.3 – Copilot produces a page

This page was more appropriate, with only one section, and even the text was useful.

Sarah then selected the **Data forms** tab, as shown in *Figure 15.4*, while still having the page selected. She had learned that it is important to have her page selected before starting a data form.

Data forms
Work with your business data

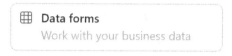

Figure 15.4 – The Copilot tab for Data forms

Copilot then asked for more details about the prompt, as shown in *Figure 15.5*. Here, Sarah could choose to use an existing table, but she let Copilot create a new table, which was prefixed with the Copilot name.

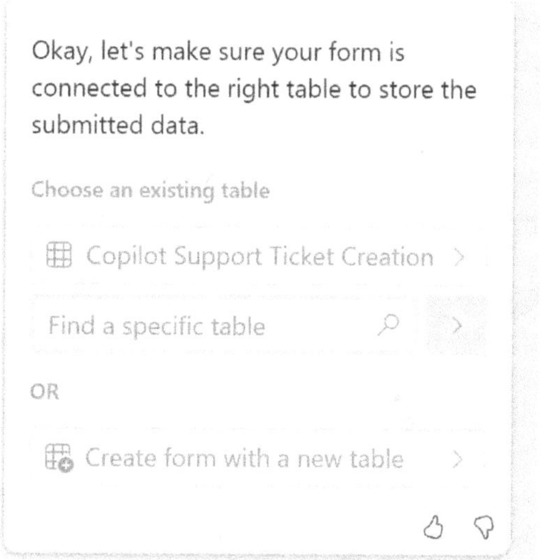

Okay, let's make sure your form is connected to the right table to store the submitted data.
Choose an existing table
Copilot Support Ticket Creation
Find a specific table
OR
Create form with a new table

Figure 15.5 – The Copilot tab for Data forms

Copilot produced a multi-step form, as shown in *Figure 15.6*. This form included sections for customer information, ticket details, and attachments. The multi-step design ensured that all necessary information was collected in an organized manner.

Figure 15.6 – The Copilot tab for Data forms

The page and form were now well-structured, allowing customers to easily submit support tickets. Sarah reviewed the form and made minor adjustments to ensure that it met the specific needs of her support system. The table created by Copilot to store the submitted tickets was functional and saved her significant time in setting up the database.

Encouraged by her success, Sarah decided to use Copilot for another task on an existing page. She wanted to create a table. Sarah selected **Sections** from Copilot and tried to create an HTML table within a section. However, this produced odd results and not the HTML table she expected, as shown in *Figure 15.7*.

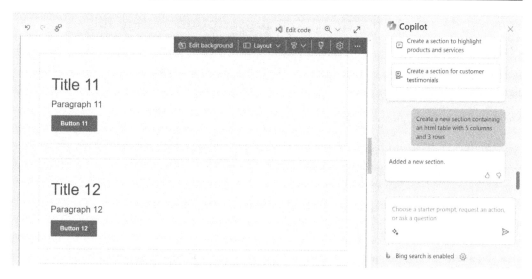

Figure 15.7 – The Copilot tab for Sections

Sarah tried to create a page with a carousel in the top section. However, this only produced a section with an image but no carousel functionality.

Sarah concluded that Copilot could create simple web pages with sections but not tables. If asked for anything else, it would either give an error message or produce a variation of an "About us" page, a customer support page, or a home page variation, but not the specific request.

Sarah also tried some design and guidance prompts, such as a detailed prompt to create subscription pages and forms to support a hosted service subscription, but Copilot produced errors. She then tried working with ChatGPT and Copilot together, and she realized that she could use ChatGPT for guidance and design ideas and Copilot to produce pages, forms, and tables.

Copilot and ChatGPT together

In this section, we'll review and outline the best approach to work with ChatGPT and Copilot together to efficiently create a customer support system. By using careful prompts, we can ensure that ChatGPT produces instructions that Copilot will understand, resulting in the correct creation of pages and forms.

To effectively use Copilot and ChatGPT together, follow these steps:

1. **Analyze customer requirements**: Begin by thoroughly understanding the customer requirements. Turn these requirements into clear and specific prompts for ChatGPT.

2. **Create clear prompts for Copilot**: Use ChatGPT to generate precise prompts for Copilot, ensuring that Copilot can produce the correct pages and forms based on the requirements.

3. **Review and tweak prompts**: After generating prompts with ChatGPT, review them carefully and tweak them as necessary to improve accuracy and ensure compatibility with Copilot.

An example scenario – creating a subscription system for a SaaS app

Feature request: Create subscription pages and forms for a SaaS app where customers can sign up, select products, and make payments.

Here are the steps for implementation:

1. **Contact registration rage**:

   - **ChatGPT prompt to Copilot**:

     - "Create a registration page with one section."
     - "Create a form for new users."

   - **Developer action**: Ensure the contact form collects essential user details

2. **Account creation page**:

   - **ChatGPT prompt to Copilot**:

     - "Create an account creation page with one section."
     - "Create a form to enter basic account details."

   - **Developer action**: Ensure the form links the user as the primary contact for the account

3. **Product selection page**:

   - **ChatGPT prompt to Copilot**:

     - "Create a product selection page with one section."
     - "Create a form with options for users to choose products and see prices."

   - **Developer action**: Ensure the page displays available products and prices correctly

ChatGPT produces a design guidance as follows.

Design and steps for the purchase subscription process

1. **Homepage button**:

   - **Action**: Add a **Subscribe to App** button on the home page

2. **Contact registration**:

   - **Action**: When the user clicks **Subscribe to App**, open the registration form created by Copilot

3. **Account creation**:

   - **Action**: After registration, navigate to the account creation page created by Copilot

4. **Product selection**:

   - **Action**: After account creation, navigate to the product selection page created by Copilot

5. **Payment integration**:

   - **Action**: After product selection, navigate to the payment page created by Copilot and handle the payment process

By following these steps, you can efficiently use ChatGPT and Copilot together to create a comprehensive subscription system for a SaaS application.

Summary

Sarah's journey as a nascent developer using Power Pages has been significantly enhanced through the integration of AI-powered tools such as ChatGPT and Copilot. These tools have not only expedited her development process but also enriched her learning experience, by providing on-demand assistance, generating code snippets, and offering debugging advice. By leveraging these tools, Sarah has been able to overcome numerous challenges, streamline her workflow, and produce high-quality, maintainable code.

In this final chapter, Sarah shared her experiences and insights on how to effectively utilize ChatGPT and Copilot in web development. She discussed the setup process, constructing effective prompts, and maintaining documentation of interactions to ensure a smooth and efficient development process. By following best practices such as iterative development, handling AI limitations, and ensuring code quality, Sarah has transformed her approach to coding, making her more productive and confident in her abilities.

One of the key ways Sarah has improved her productivity is by asking ChatGPT for snippets of code for JavaScript, Liquid, and design, as well as implementation ideas. This includes seeking step-by-step guidance to ensure that she is on the right track. By doing so, she has been able to quickly incorporate necessary functionalities and design elements into her projects.

AI tools have proven to be invaluable companions in Sarah's journey, providing real-time collaboration, reducing cognitive load, and improving the overall quality of her projects. Through detailed examples and practical tips, Sarah illustrated how developers can harness the power of AI to enhance their skills and streamline their development workflows. As she reflects on her progress, Sarah recognizes that the integration of AI tools has been a game-changer, paving the way for a more efficient, enjoyable, and productive coding experience.

By using ChatGPT and Copilot, Sarah has not only accelerated her development process but also gained a deeper understanding of coding practices and improved the quality of her projects. These tools have empowered her to tackle more complex tasks with confidence, ultimately transforming her approach to web development.

Index

packtpub.com

Subscribe to our online digital library for full access to over 7,000 books and videos, as well as industry leading tools to help you plan your personal development and advance your career. For more information, please visit our website.

Why subscribe?

- Spend less time learning and more time coding with practical eBooks and Videos from over 4,000 industry professionals

- Improve your learning with Skill Plans built especially for you

- Get a free eBook or video every month

- Fully searchable for easy access to vital information

- Copy and paste, print, and bookmark content

Did you know that Packt offers eBook versions of every book published, with PDF and ePub files available? You can upgrade to the eBook version at packtpub.com and as a print book customer, you are entitled to a discount on the eBook copy. Get in touch with us at customercare@packtpub.com for more details.

At www.packtpub.com, you can also read a collection of free technical articles, sign up for a range of free newsletters, and receive exclusive discounts and offers on Packt books and eBooks.

Other Books You May Enjoy

If you enjoyed this book, you may be interested in these other books by Packt:

Learn Microsoft Power Apps

Matthew Weston, Elisa Bárcena Martín

ISBN: 978-1-80107-064-5

- Understand the Power Apps ecosystem and licensing
- Take your first steps building canvas apps
- Develop apps using intermediate techniques such as the barcode scanner and GPS controls
- Explore new connectors to integrate tools across the Power Platform
- Store data in Dataverse using model-driven apps
- Discover the best practices for building apps cleanly and effectively
- Use AI for app development with AI Builder and Copilot

Low-Code Application Development with Appian

Stefan Helzle

ISBN: 978-1-80020-562-8

- Use Appian Quick Apps to solve the most urgent business challenges
- Leverage Appian s low-code functionalities to enable faster digital innovation in your organization
- Model business data, Appian records, and processes
- Perform UX discovery and UI building in Appian
- Connect to other systems with Appian Integrations and Web APIs
- Work with Appian expressions, data querying, and constants

Packt is searching for authors like you

If you're interested in becoming an author for Packt, please visit authors.packtpub.com and apply today. We have worked with thousands of developers and tech professionals, just like you, to help them share their insight with the global tech community. You can make a general application, apply for a specific hot topic that we are recruiting an author for, or submit your own idea.

Share Your Thoughts

Now you've finished *Microsoft Power Pages in Action*, we'd love to hear your thoughts! Scan the QR code below to go straight to the Amazon review page for this book and share your feedback or leave a review on the site that you purchased it from.

https://packt.link/r/1837630453

Your review is important to us and the tech community and will help us make sure we're delivering excellent quality content.

Download a free PDF copy of this book

Thanks for purchasing this book!

Do you like to read on the go but are unable to carry your print books everywhere?

Is your eBook purchase not compatible with the device of your choice?

Don't worry, now with every Packt book you get a DRM-free PDF version of that book at no cost.

Read anywhere, any place, on any device. Search, copy, and paste code from your favorite technical books directly into your application.

The perks don't stop there, you can get exclusive access to discounts, newsletters, and great free content in your inbox daily

Follow these simple steps to get the benefits:

1. Scan the QR code or visit the link below

https://packt.link/free-ebook/9781837630455

2. Submit your proof of purchase
3. That's it! We'll send your free PDF and other benefits to your email directly